,,,

*Devoted to immortality, perhaps
the only more ultimate quest.*

100 GREAT PERPETUAL MOTION MACHINES AND SIMILAR INVENTIONS

ARRANGED ALPHABETICALLY

by Nathan Coppedge

PREFACE / INTRO

This is what may be the first nearly complete tome of working free energy inventions. 50 Great Flying and Underwater Perpetual Motion Machines are saved for a second volume. Nathan Coppedge is an inventor like no other. Most of these creations are of Nathan's own invention, replete with simple as-easy-as-it-gets mathematics. Many of the diagrams provide critical information on how to possibly build the devices and make them work at least as partial demonstrations.

,,,

1st Fully Provable

WHAT I CALL THE "1ST FULLY PROVABLE PERPETUAL MOTION MACHINE"

[MODIFIED FOR GREATER WORKABILITY]   DIAGRAM WITH STATS BY

FEB 14, 2021

9 degree angle is about 58.25 application. Max lvg is 2.24, Min is 1.73.
Min HvyMass = Max Lvg X 0.5825 + 1 = 2.3 mass units, (first check if able to lift ball to max leverage)
Max HvyMass = Min Lvg + 1 = 2.73 mass units, (check if still able to lift ball at min leverage)
Then there is a window of nearly 0.43 the mass of the marble
to account for friction! In a balance! So, it works!

weight ratio: as one quarter,
one penny 1 marble and 5
in of duct tape is to 1 marble
lever ratio 10.8 - 14 : 6.25
track angle ~0.5deg
upwards-sloped
lever angle
est 2 - 9 deg
downwards-
sloped.

NOTE: NEW LIP TO
DIRECT FREE-FALL
MAY BE IMPORTANT

NATHAN LARKIN COPPEDGE

1.   **2.**   **3.**

B.1   A 1   A 2   B. 2

STEP 1: BALL HAS ALTITUDE
TO APPLY PRESSURE TO
FIRST MODULAR
COUNTER-WEIGHTED
LEVER. (B.1)

WORKING RATIOS in straight lever experiment.

STEP 2: BALL RISES ALONG INCLINE,
HELPED BY FIXED TRACK SUPPORT A1.
UNTIL IT REACHES BEGINNING OF
2ND MODULAR LEVER. (B.2)
STEP 3: BALL WEIGHT HAS SUFFICIENT
HEIGHT TO ACTIVATE LEVER AT SAME
INITIAL HEIGHT, IN SPITE OF ITS
LOWER BASE HEIGHT. PMM!

- Note: There is also a Reverse 'R'
  version of the device (R1FP) using the
  same ratios and same main
  components that directs the ball
  inwards using a lower fulcrum height,
  with the lever being depressed on the
  outer end before upward motion. That
  device was the source of the first
  evidence of a self-powered flying

machine which had no net inputted energy.

- Date of Invention: July 12, 2016.
- State of affairs: Strong suggestion of workability. Not easy due to modular construction and in some cases need for extremely heavy weights or perfect hollow metal spheres.
- <131% conventional Over-Unity.
- Leverage: 1.75-2.25: 1
- Counterweight Mass: >2.125 to <2.75X (previous estimate 1.8X to 2.4X)
- Maximum Gradient: Approx < 13.05 degrees (not calculated).

1st Successful Module Perpetual Motion ("Cat-Trap")

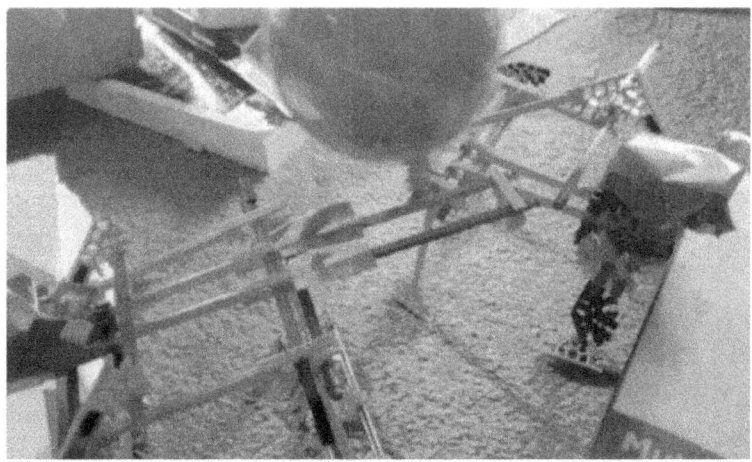

- Date of Invention: November 12, 2020.
- Note: Involves some unique innovations, in particular the blocking for the lever on the lower right is slightly higher than the blocking on the far lower left due to the longer lever on the left, although both lower block points are slightly (perhaps visibly) below the high point of the fulcrum pin.
- State of affairs: In process.
- Rating: <150% conventional Over-Unity.

- Leverage: 1.25:1 (judging by counterweight distance)
- Counterweight Mass: >1.625X to <2.25X (assumes 1X additional weight in end that is used as a lever. Practical ideal average about 1.25 surprisingly, since lever end does not always require much extra mass).
- Maximum Gradient: < 22.5 degrees.

Amish Perpetual Motion, Improved

IMPROVED AMISH OVER-UNITY

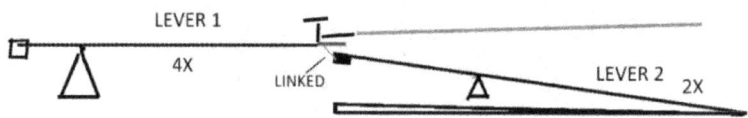

LEVER 1
4X
LINKED
LEVER 2 2X

>1X to compensate
for structural mass
and to hit stopper

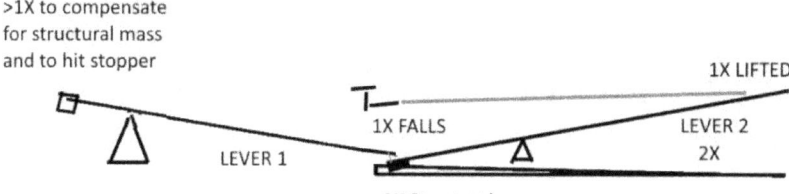

1X LIFTED

1X FALLS
LEVER 2
2X
LEVER 1

2X Structural
Mass Advantage

STEP 1:
LEVER 1 RISES USING COUNTERWEIGHT MASS,
LIFTING LIGHTWEIGHT STOPPER, SENDING BALL
DOWNWARDS, WITH LITTLE TO NO RESISTANCE FROM LEVER 1
DUE TO HIGH LEVERAGE RATIO OF 4X (STRUCTURAL MASS ASSISTS).

PRESSURE FROM BALL IS SENT THROUGH LEVER 2 ON SHORT END,
ASSISTED BY 2X STRUCTURAL MASS ADVANTAGE ON SHORT END,
BRINGING EQUIVALENT OF 3X MASS IN RATIO OF 1:2,

LIFTING ONE OF A SERIES OF BALLS PREPARED TO ROLL ALONG
A RAMP AT THE BASE OF LEVER 2, USING OPPOSITE END OF LEVER 2.

1X BALL IS DEPOSITED AT BASE OF LEVER 2, WHERE IT MAY ROLL.

MARBLE THAT IS LIFTED JOINS UPPER RAMP, ROLLING TO
JOIN THE SERIES OF MARBLES AT THE BEGINNING.

BALL RETURNS TO LEVER 1

LEVER 1 COUNTERACTS STRUCTURAL MASS OF 1X
AND HAS ADDITIONAL PRESSURE TO LIFT LIGHTWEIGHT STOPPER,
INCREMENTING NEXT BALL.

2023-03-05

Anachronistic Ground Transport

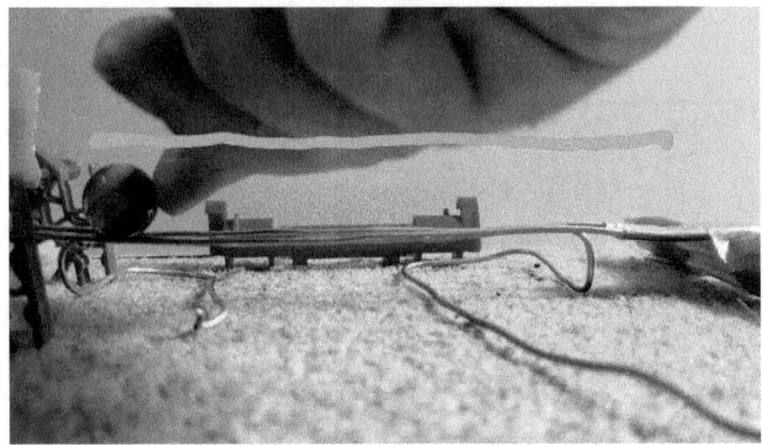

Larger, quieter transportation (huge free energy train-buildings could house other vehicles which move faster inside the larger vehicle!).

Apollo Device 2

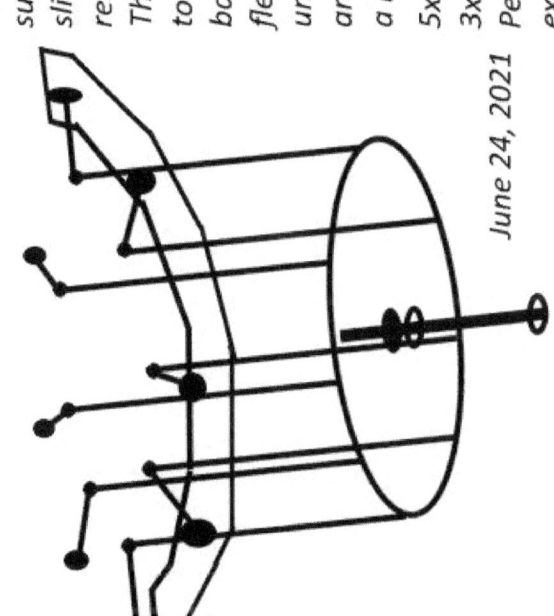

APOLLO DEVICE 2 Nathan Coppedge

The rising weights are supported on a very slight upwards incline relative to the hinges. The hinges are blocked to be perpendicular to the bar, but allow upward flexion of the bar. The unsupported members are blocked, so provide a kind of leverage.

5x 0.5 equals 2.5

3x 1 equals 3

Perpetual motion, when extremely low friction!

June 24, 2021

With 5 arms, 3 X 0.5 equals 1.5, 2 X 1 equals 2. 2 - 1.5 equals 0.5 plus 1 equals 150% OU.

Lvg: 1:0.5

Ratio: 1:1

Over-Unity: < 150% (?)

Bar Crescent Device

(Pictured Middle)

- Date of Invention: By August 14, 2019.
- Secretive clues: By December 29, 2016.
- State of affairs: Not tested.
- Rating: < 125% conventional Over-Unity.
- Lvg: 1.5–2:1
- Mass: > 2 to < 2.5
- Maximum Gradient: < 13.05 degrees.

Bobblety Toy

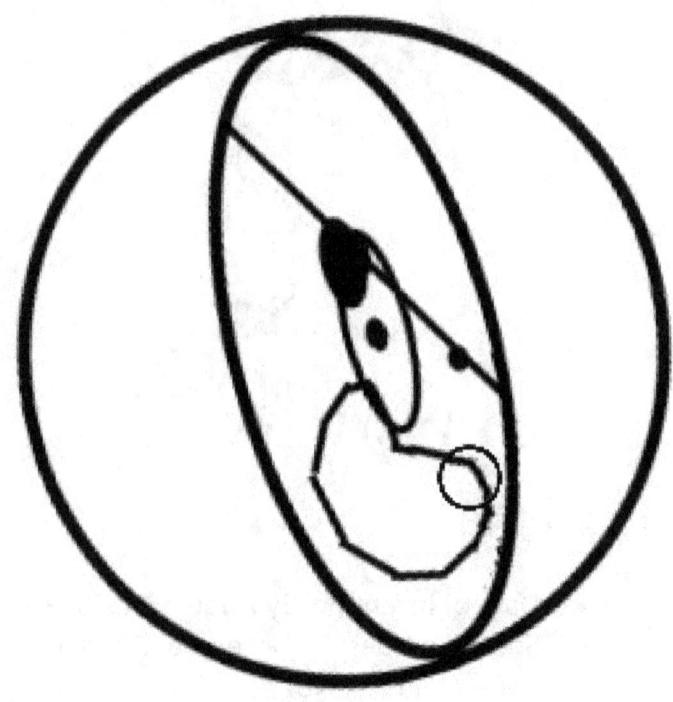

Bridge 2

PERPETUAL MOTION BRIDGE 2
MECHANICAL ENGINEERING CONCEPT

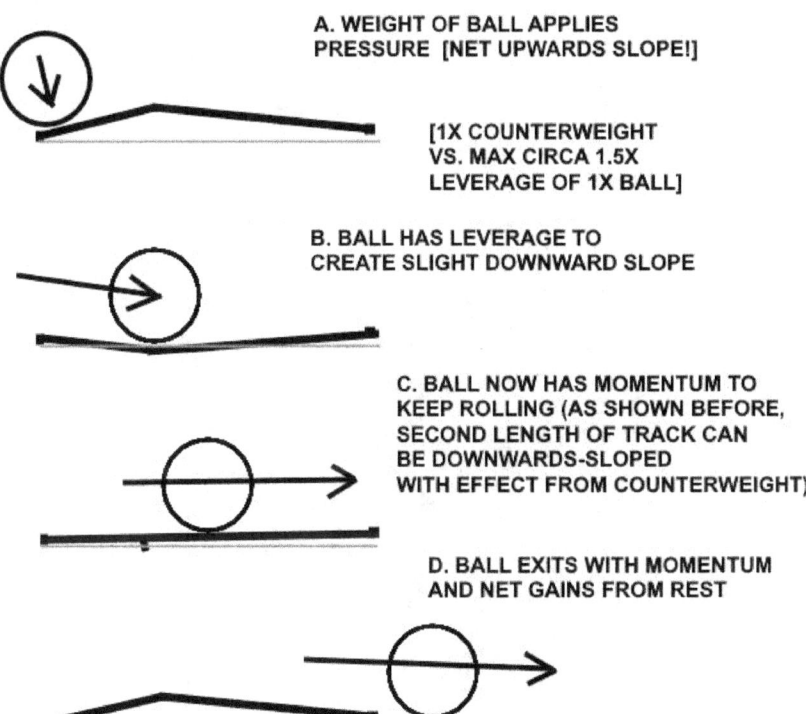

A. WEIGHT OF BALL APPLIES
PRESSURE [NET UPWARDS SLOPE!]

[1X COUNTERWEIGHT
VS. MAX CIRCA 1.5X
LEVERAGE OF 1X BALL]

B. BALL HAS LEVERAGE TO
CREATE SLIGHT DOWNWARD SLOPE

C. BALL NOW HAS MOMENTUM TO
KEEP ROLLING (AS SHOWN BEFORE,
SECOND LENGTH OF TRACK CAN
BE DOWNWARDS-SLOPED
WITH EFFECT FROM COUNTERWEIGHT)

D. BALL EXITS WITH MOMENTUM
AND NET GAINS FROM REST

Carbonation Swivel

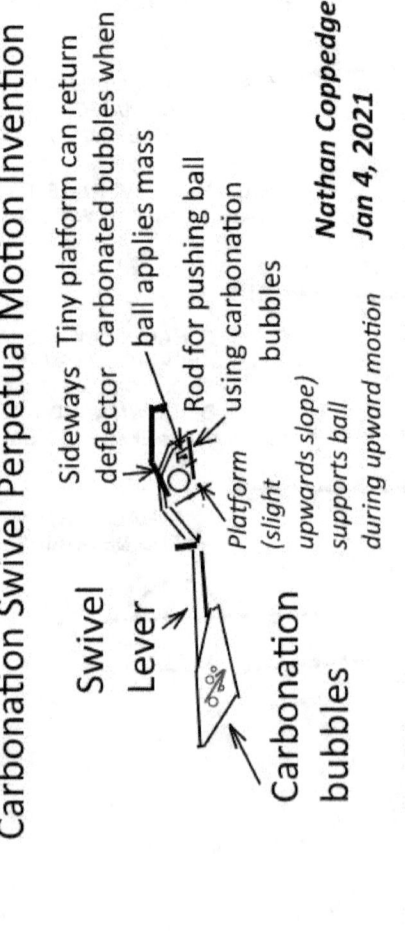

Carbonation Swivel Perpetual Motion Invention

Sideways Tiny platform can return
deflector   carbonated bubbles when
ball applies mass

Rod for pushing ball

using carbonation
bubbles

*Platform
(slight
upwards slope)
supports ball
during upward motion*

Swivel

Lever

Carbonation
bubbles

*Nathan Coppedge
Jan 4, 2021*

Motion of bubbles is upward and counter-clockwise, as shown by grey arrow

Casimir Device

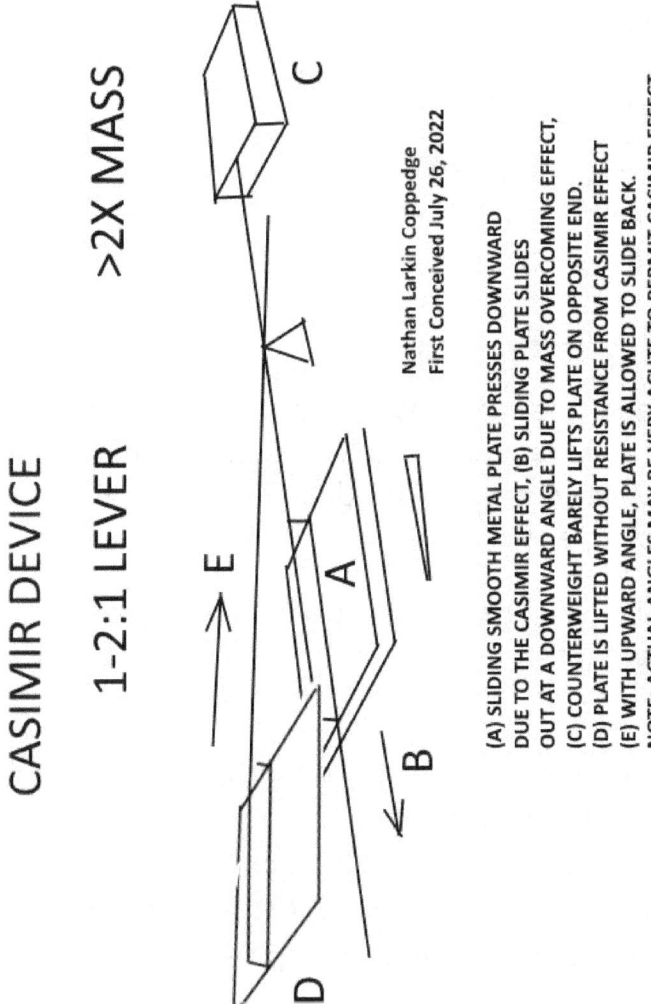

CASIMIR DEVICE

1-2:1 LEVER

>2X MASS

Nathan Larkin Coppedge
First Conceived July 26, 2022

(A) SLIDING SMOOTH METAL PLATE PRESSES DOWNWARD
DUE TO THE CASIMIR EFFECT, (B) SLIDING PLATE SLIDES
OUT AT A DOWNWARD ANGLE DUE TO MASS OVERCOMING EFFECT,
(C) COUNTERWEIGHT BARELY LIFTS PLATE ON OPPOSITE END.
(D) PLATE IS LIFTED WITHOUT RESISTANCE FROM CASIMIR EFFECT
(E) WITH UPWARD ANGLE, PLATE IS ALLOWED TO SLIDE BACK.
NOTE: ACTUAL ANGLES MAY BE VERY ACUTE TO PERMIT CASIMIR EFFECT.

Chemical Perpetual Motion

April 27, 2019

Maybe a chemical grid or chemical storage.

This would likely work with electricity.

It is not likely to take off unless chemicals are cheaper and more efficient than mechanical perpetual motion.

I mean this literally.

For example, the current version of chemical storage is gas or batteries, but gas is fuel-based, which is to say it requires ongoing expenditures, and batteries don't generate at all.

Chemical or semi-nano perpetual motion might be 500+ years in the future even if mechanical perpetual motion takes off now.

Before I knew much of what to do with the function spectrum, my mind leapt to a possibility with Difference +5. My immediate thought was that this value, which represented < 500% OU corresponded with some type of chemical development in perpetual motion. There is thought to be some way in which magnets, or perhaps separated compartments of liquid, can interact, using a perhaps newly-discovered mechanical chemistry which adopts some of the properties of perpetual motion.

Chinese Curve Rail, Improved

MIN FALLING RATIO FOR MOTION = 0.59375

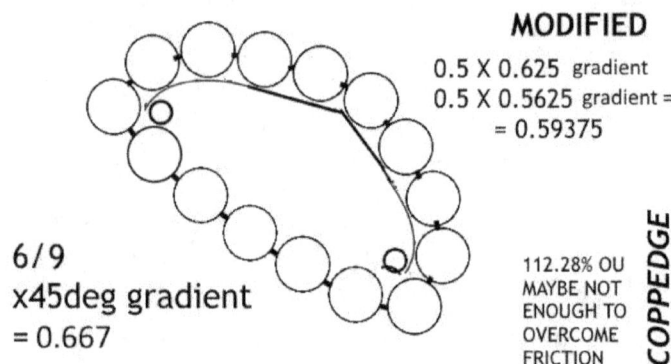

MODIFIED

0.5 X 0.625 gradient
0.5 X 0.5625 gradient =
= 0.59375

6/9
x45deg gradient
= 0.667

112.28% OU
MAYBE NOT
ENOUGH TO
OVERCOME
FRICTION

NATHAN COPPEDGE

ACTUAL FALLING RATIO: 0.667

FUNCTIONAL?
CHINESE PATENT: CN105587479A

- NOTES: Alternately, buoyancy rail (inverted, with buoys).
- Date of Invention: 2015 (Chinese)
- Possible precedents: Nathan Coppedge (2006).
- Note: This device is assumed to be dysfunctional, however the diagram may suggest over-unity.
- State of affairs: Now thought to be dysfunctional when friction is included

- Rating: < 112.28% conventional Over-Unity.
- Ratio of Falling Mass: 0.7058
- Ratio of Rising Mass (Non-Conservative): 0.59375
- Falling Gradient: < 45 degrees.

Compensation Lever Device

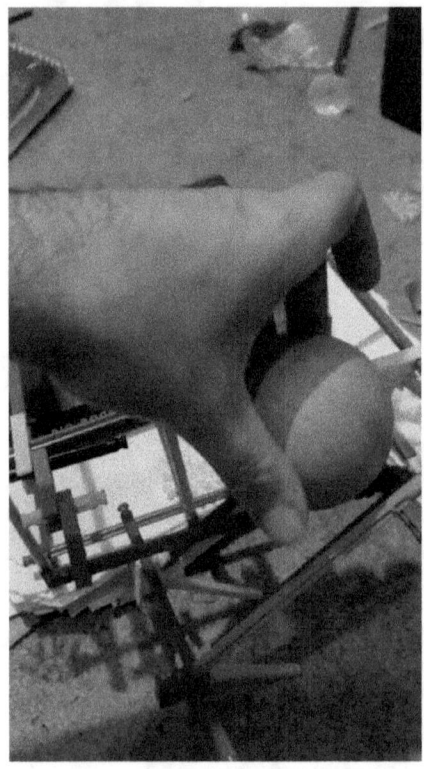

- Date of Invention: January 29, 2020.
- Precedents: Similar work from 2009
- State of affairs: Some weak partial experimental evidence.
- Rating: < 123.5% Conventional Over-Unity.

- Leverage: 1: 1.5 (range could actually be > 1:1 < 1:2, 1:1.5 is simply close to the best choice)
- Ball Mass: >1.75 < 2.5 (Note: This value was previously given as the counterweight's mass, with the leverage ratio of 1:1.5. This was incorrect. With a leverage ratio of 1: 1.5 with 1X being the side depicted, it must be the ball which has a value of >1.75 to <2.5. The counterweight by comparison has a mass of 1X, with an additional 1X incorporated in the structure of the longer end opposing the end with the ball which is not included in the counterweight's mass).

Computer Applications

Several theories:
* Essential formula initially, is that the essence is energy, performing functions.
* Supplement is energy engages processes activated by energy.
* Further is efficiency activated by multiple processes.
* Exponential efficiency is key in this conception.
* Now, what we call conventional processes, which could be regular computing or exponentially
efficient *computing*. This forms the 3-layer structure.
* Now, treat all that as the initial premise.
* Imagine it could be modified to supply the hardware.
* Imagine distinct linkages of this process could be micro components.
* Now we get concepts like fast overprocessing, bulk loading, physical layer conversion, and constant overhead (overhead overdrive), which may lead to advances.
* More distinctly, it is the theme of utility computers and permanent computers, energy-enhanced hand-utilities and wire computers.

See also under Perpetual Motion Mathematics and Reactive Mechanisms, and see following page. Also note perpetual motion has sometimes been supported by an A.I.

("OVER-UNITY WIRES") OVER-UNITY WAVELENGTH CONDUCTORS THEORETICAL CONCEPT -- NOT KNOWN IF TESTED

HIGH-WAVELENGTH-PRIMARY DEVICES

CONCEPT	HIWAV	LOWWAV/AVMINMAX	RATING
GENIE	1:1	Limited Resist. Mod	<200%
BRIDGE2	1.5:1	>1.75 <2.5	<162.5%
MBC	2.7/6	Resistance	<155%
RL 4.2	5:1	>2.5 <6	<150%
IVL	4:1	>3 <5	<150%
VLVR	3:1	>2.5 <4	<150%
NIBW4	3:1	>2.5 <4	<150%
SWNL	3:1	>2.5 <4	<150%
BRCKL	3:1	>2.5 <4	<150%
VIM1	3:1	>2.5 <4	<150%
SPLGRA	2.5:1	>2.25 <3.5	<150%
SMVL	2:1	>2 <3	<150%
SPRLCO	2:1	>2 <3	<150%
VIM2	2:1	>2 <3	<150%
NAL	2:1	>2 <3	<150%
BRSCIS	2:1	>2 <3	<150%
TORQ360	1.5:1	>1.75 <2.5	<150%
Cat-Trap	1.25:1	>1.625 <2.25	<150%
TLTMTR	1:0.5 CON2	>1 <2	<150%
SlaP-W	1:0.5	>1 <2	<150%
HSwrW	1:0.5	>1 <2	<150%
D-DD	1:0.5	>1 <2	<150%
APOLLO2	1:0.5	5 ARMS	<150%
MACHINE	**LVG**	**MIN*	MAX***	**RATING**

CONCEPT	HIWAV	LOWWAV/AVMINMAX	RATING
EschDelta	1:0.5	BALL 1X NO CW	<150%
HptSymVL	1:1	>1.5 <2	<150%
APOLLO1	1:0.5	RATIO 0.75	<150%
ELNMSK	1:2.5	BALL>2.25 <3.5	<140%
MAEsxh		+<20% Magnetism	<<140%
I-CPM	1:1	>0.6875 <1.375 Avg	<137.5%
*RGHLVR	2.5-3:1	>2.5 <3.5	<136% AVG
*LGLVR	2.5-3:1	>2.5 <3.5	<136% AVG
ISTFP	1.75-2.25:1	>2.125 <2.75	<131% AVG
ESCHLVR	3-4:1	>3 <4	<129% AVG
iSLD	3-4:1	>3 <4	<128.6% AVG
BarCr	1.5:2:1	>2 <2.5	<125%
2SVLW	2:1	8 ARMS	<125%
TrpLD	1.5-2:1	>2 <2.5	<125%
SwivBal	1:1	>2.5 <3 (/1 Arm)	<125% DIFF 2
MACHINE	**LVG**	**MIN*	MAX***	**RATING**

R1FP	2-3:1	>2.5 <3	<120%
CRSC	2-3:1	>2.5 <3	<120%
3RL	2-3:1	2-5 <3	<120%
A-GDB2	2-3 :1	>2.5 <3	<120%
MACHINE	**LVG**	**MIN*	MAX***	**RATING**

CONCEPT	HIWAV	LOWWAV/AVMINMAX	RATING
V SLANT	2-3:1	>2.5 <3	<120%
Coquet	2-3:1	>2.5 <3	<120%
SCoget	2-3:1	>2.5 <3	<120%
SpiPLD	1-1.5:1	BALL>1.75<2	<120%
PalivrL	1-1.5:1	BALL>1.75<2	<120%
HIN	1-1.5:1	>1.75 <2	<120%
NIBW1.5	1-1.5:1	>1.75 <2	<120%
VNIBW5	1-1.5:1	>1.75 <2	<120%
NIBW 6	1-2:3	BALL>2.5 <4	<119.23%
JBUBL	1-1.5:1	BUB >2-75<3	<116.67% 2 DIFF
HeBall	1:0.5	Mass >1 <2	<116.5%
HeBall2	1:0.5	Bycy >1 <2	<116.5%
BiAp	1:4	Ball >3 <5	<115.56% (T)
R1FP	1.25-2:1	>2 < 2.25	<115.4% (T)
CurvR	Gravonly	0.7058/0.58375 Resistance	<112.28%
IWW	0.94736 :1.0453	Resistance	<110.34%
RepSpi	1.25:1	>0.625 < 1.25 (Equil)	<110.25%
BstBycChn	3-root of 22.5 deg		<102.82%
MiniWaterwheel	Moving parts: 2		Unity+
ScarpP	Moving parts: 2		100% +/-
GravMot	Moving parts: 1		<100%
FakeMagnet	Moving parts: ?		<100%
Grav-B2	Moving parts: 63		<20%
THConv	Moving parts: 25		<7.33%
ConvWheel	Moving parts: 17		<1%
MACHINE	**LVG**	**MIN*	MAX***	**RATING**

SELECT LOW WAVELENGTH-PRIMARY DEVICES (IF RELEVANT)

CONCEPT	LOWWAV	HIWAV/AVMINMAX	RATING
RSWIV	1:1.5	BALL>1.75 <2.5	<150%
DSSMM	1:0.5	BALL >1 < 2	<150%
SwaP	1:0.5	(1:1)	<150%
EschDelta	1:0.5	BALL 1X N0 CW	<150%
A-GDB1	1.25:1	>0.625 <1.25	<150%
INIBW2	11-13:1	7.5 TO 12. wedge	<148.75%
ModBD1	1.75-2-25.1	>1.125<1.75	<133.75% -1 DIFF
COMPL	1:1.5	BALL >1.75 <2.5	<123.5%
DROPLV	1:2.3	BALL>2.5 <4	<123.08% AVG
ESCH	1:0.5	BALL 1X NO CW	<120%

SOME EARLIER NOTES RETAINED

EQUATIONS:

Min Heavier Mass = (Max Lvg / 2) + 1

Max Heavier Mass = Min Lvg + 1

Min Lvg = Max Heavier Mass - 1

Max Lvg = (Min Heavier Mass - 1) X 2

Over-Unity = Heavier Mass Rng / Lvg Ratio + 1 X 100 (%)

Smaller Mass = 1X

PM Cars Extra Mass < OU - 100%

Flying Vehicles Extra Mass < OU - 200%

Flying does not work when

Planetoids: Phi / 2 + 1 * 100

Lvg Rng >= 1/2 max leverage.

= < 130-9% Conventional Over-Unity

* (min/max of CW unless stated as ball)

OVER-U CALC = Mass Rng / Lvg Ratio + 1 X 100 (%)

Normally, Ball has Mass = 1

Min / Max refers to ctrweight unless stated otherwise. Unless otherwise stated, the ctrweight, which is typically heavier, is located on the shorter end of the lever. In some cases a lever is not used, and alternate formulas may apply. Important (!): If lever is present... Long end of lever additionally = <1 but >0 constant. Most make use of the properties of a balance

NATHAN LARKIN COPPEDGE

Coquette

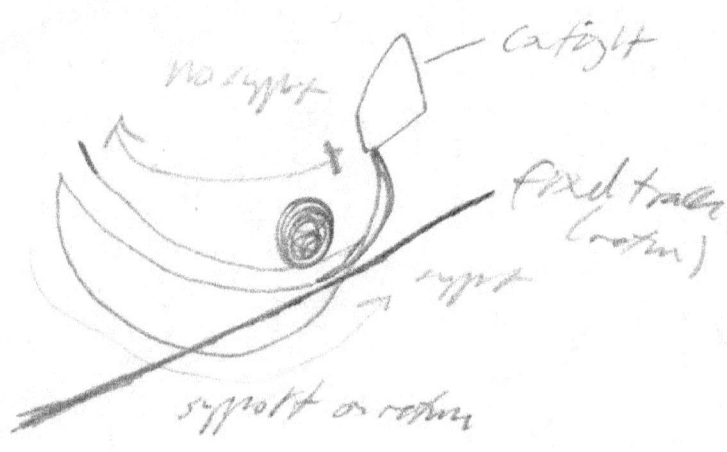

- Date of Invention: April 16, 2009
- State of affairs: Thought to work with the side support which may require additional design.
- Rating: < 120% conventional Over-Unity.
- Leverage: 2 to 3 : 1
- Counterweight Mass = >2.5 <3
- Maximum Gradient: < 9 degrees.

Corkscrew Belt

CORKSCREW BELT
NCOPPEDGE

Crescent Leverage Device

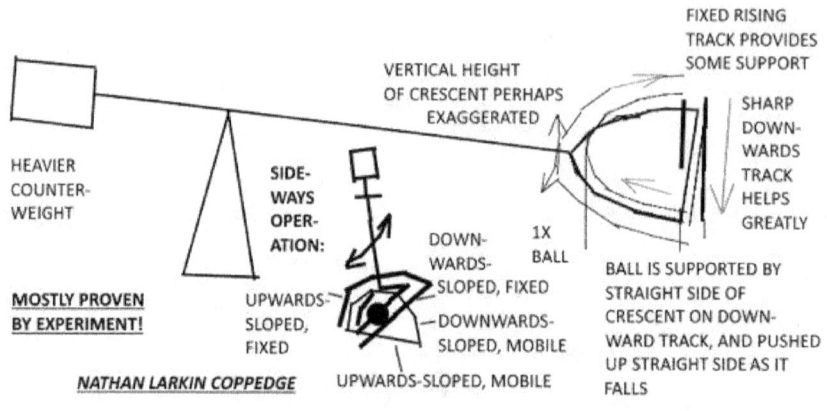

CRESCENT LEVER PERPETUAL MOTION DEVICE   <MODIFICATION>

- Date of First Evidence: July 24, 2018.
- State of affairs: Some collaborations, but little progress. An experiment shows theoretical workability. Likely depends on a lightweight design which may require adjusting counterweight ratio to compensate for required structural mass.
- Rating: < 120% conventional Over-Unity.
- Leverage = 2 to 3:1
- Counterweight Mass = >2.5X to <3X
- Maximum Gradient: < 9 degrees.

Csaba Horvath Principle

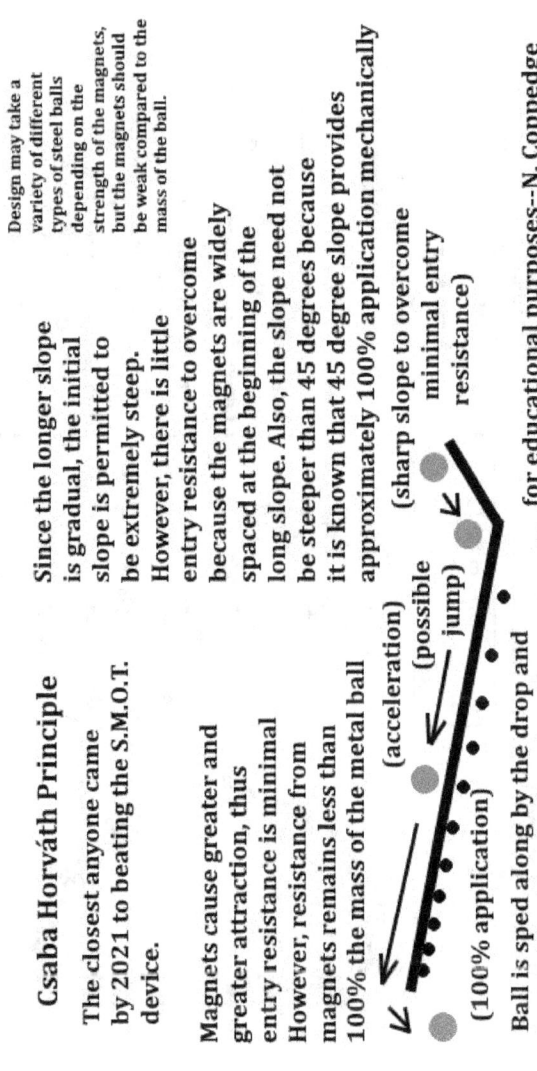

Csaba Horváth Principle

The closest anyone came by 2021 to beating the S.M.O.T. device.

Magnets cause greater and greater attraction, thus entry resistance is minimal. However, resistance from magnets remains less than 100% the mass of the metal ball

(100% application)

Ball is sped along by the drop and by initial wide spacing of weak magnets (?)

Since the longer slope is gradual, the initial slope is permitted to be extremely steep. However, there is little entry resistance to overcome because the magnets are widely spaced at the beginning of the long slope. Also, the slope need not be steeper than 45 degrees because it is known that 45 degree slope provides approximately 100% application mechanically

(acceleration)

(possible jump)

(sharp slope to overcome minimal entry resistance)

Design may take a variety of different types of steel balls depending on the strength of the magnets, but the magnets should be weak compared to the mass of the ball.

for educational purposes--N. Coppedge

31

Curve Rail, Early

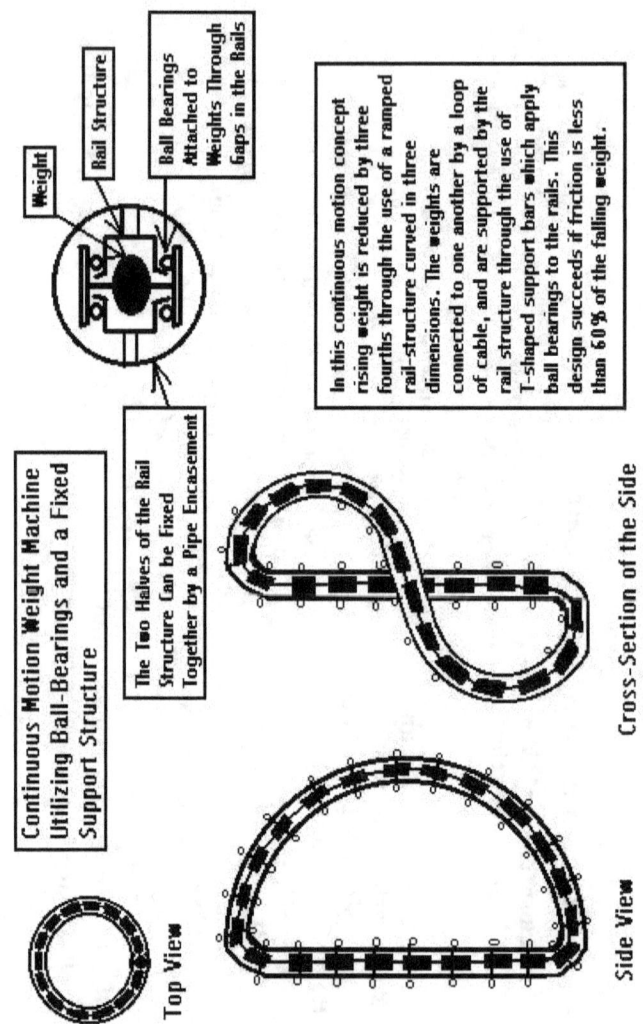

Weight

Rail Structure

Ball Bearings Attached to Weights Through Gaps in the Rails

In this continuous motion concept rising weight is reduced by three fourths through the use of a ramped rail-structure curved in three dimensions. The weights are connected to one another by a loop of cable, and are supported by the rail structure through the use of T-shaped support bars which apply ball bearings to the rails. This design succeeds if friction is less than 60% of the falling weight.

Continuous Motion Weight Machine Utilizing Ball-Bearings and a Fixed Support Structure

The Two Halves of the Rail Structure Can be Fixed Together by a Pipe Encasement

Cross-Section of the Side

Top View

Side View

Dominoes of Increasing Heights

...

Dominoes. The Original Self-Resetting

PERPETUAL MOTION DOMINOES ITERATION 2

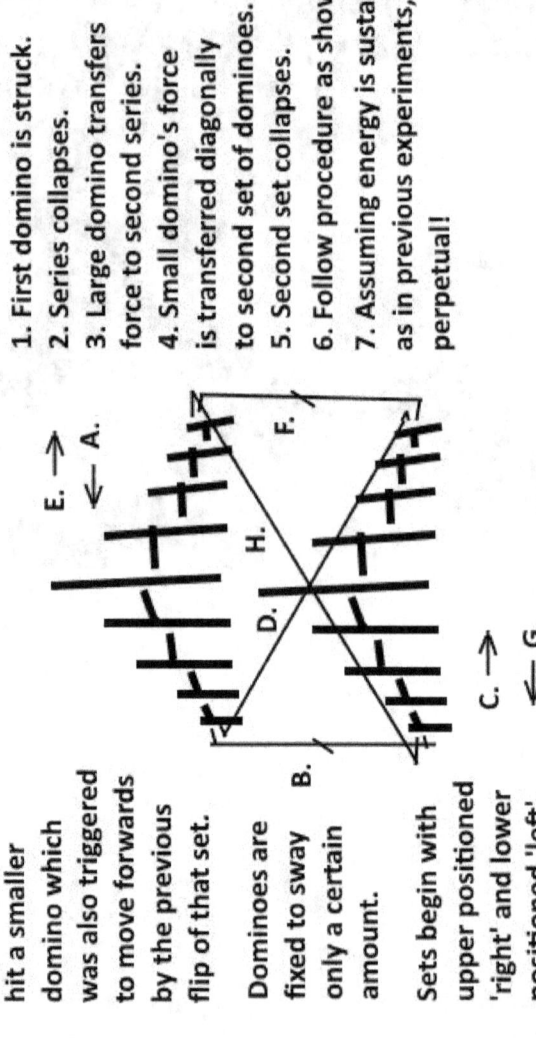

E. → ← A.

H.

D.

F.

B.

C. → ← G.

1. First domino is struck.
2. Series collapses.
3. Large domino transfers force to second series.
4. Small domino's force is transferred diagonally to second set of dominoes.
5. Second set collapses.
6. Follow procedure as shown.
7. Assuming energy is sustained as in previous experiments, perpetual!

Diagonal ones hit a smaller domino which was also triggered to move forwards by the previous flip of that set.

Dominoes are fixed to sway only a certain amount.

Sets begin with upper positioned 'right' and lower positioned 'left'.

Nathan Larkin Coppedge   March 12, 2023

Double-Disk Device

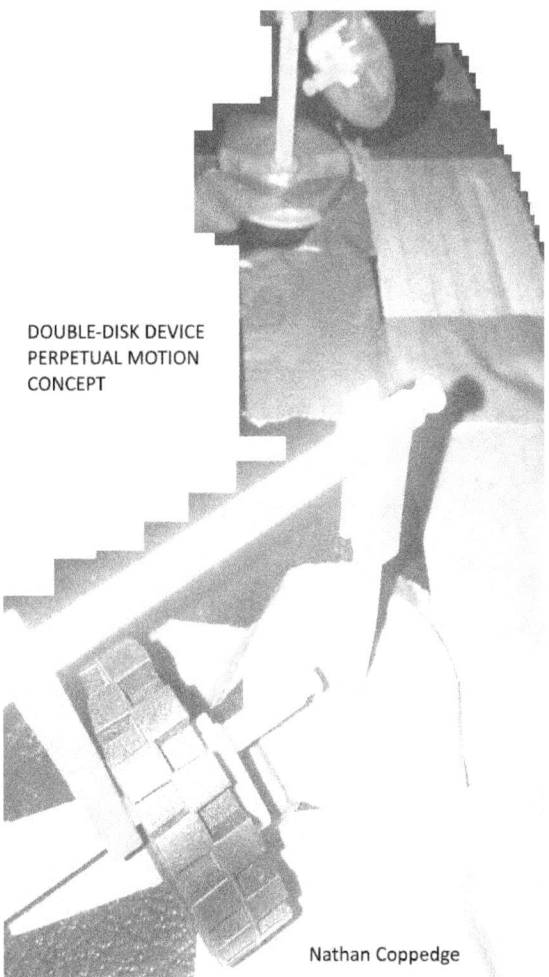

DOUBLE-DISK DEVICE
PERPETUAL MOTION
CONCEPT

Nathan Coppedge

- First Attempted Demonstration: Aug 11, 2018.

- State of affairs: May require additional cleverness.
- Rating: < 150% conventional Over-Unity
- Leverage: 1:0.5
- Counterweight Mass: (1:1)
- Maximum Gradient: < 22.5 degrees.

Double-Lever

Key discovery recorded by Nathan Coppedge February 6, 2021:

Keep in mind, this may be the best antigravity device ever created. Depending on the size of the apparatus, the heavy weight could be lifted to very high altitudes merely by rotating a very thin, lightweight lever. This is the ultimate of efficiency for causing heavy weights to be lifted at very low energy cost.

Drop Lever / Outer Swivel Device

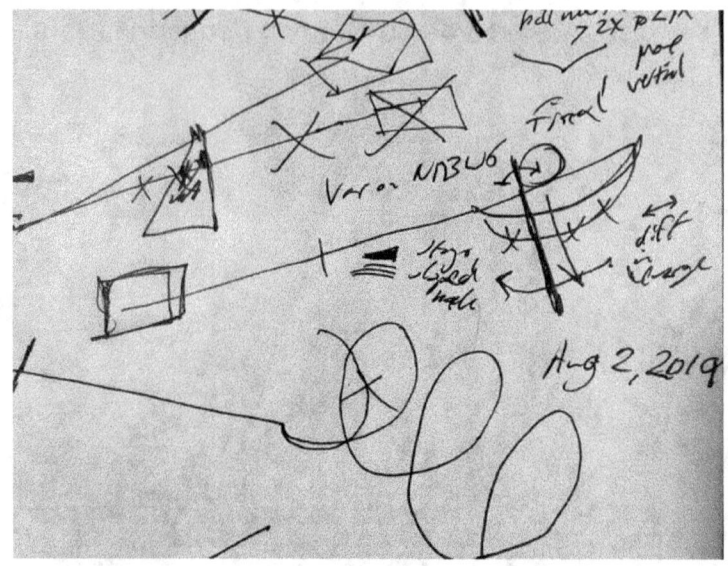

- Date of Invention: August 2, 2019.
- State of affairs: An intriguing design that may be worth attempting. Some frustrations expected, and some applications.
- Rating: Adjusted rating in terms of weight of ball < 123.08% (1.5 / 2 + 1 X 100 AVG) < 175% in terms of 1X mass.
- Leverage: 1-2:3
- Ball Mass: >2.5X to <4X

Dual-Seesaw Device Motive Mass Machine

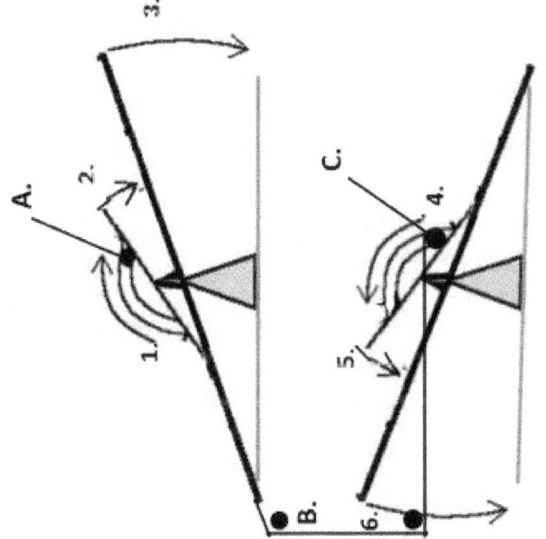

**MOTIVE MASS PERPETUAL MOTION
ITERATION 3**

Part 1.
Ball weight A. is massive enough that when moved upwards, it causes motions 1, 2, 3. Motion 3 is transferred through Pulley System B.

Part 2.
C. rises a short distance (permitted by the long distance of 3, e.g. because 3 acts as a lever, and the track beneath C. acts as a wedge, creating motions 4, 5, 6.

Overall: If the above is deemed sufficient for a self-sustaining cycle, a Motive Mass Machine could be created.

Dec 24, 2015 --- Nathan Coppedge

- Date of Invention: Dec 24, 2015.
- State of affairs: May require flatter trajectory.
- Rating: < 150% conventional Over-Unity
- Leverage = 1:0.5 (Unconventional Uses Limited Resistance Chain Reaction)
- Maximum Gradient: < 22.5 degrees.

Enhanced Energy

- Enhanced energy technologies (someone may discover the secret of enhancing energy from perpetual motion with other perpetual motion machines, creating more energy in exponential proportion to electric output).

Escher Delta

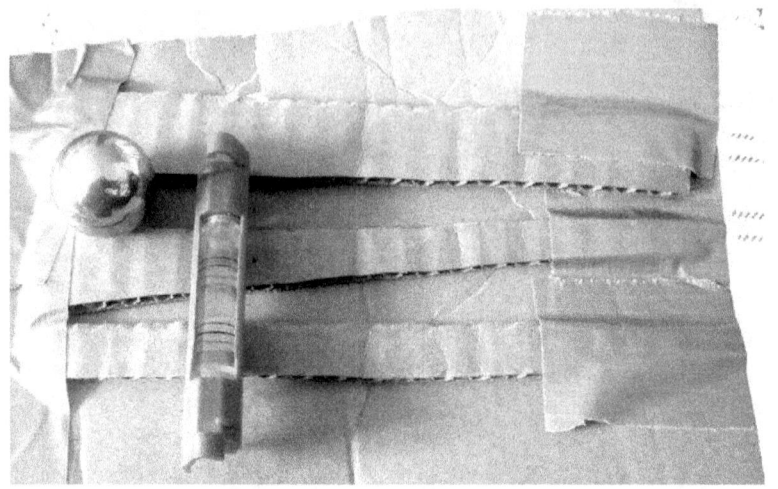

- Date of First Evidence: September 11, 2018.
- State of affairs: Compelling partial evidence, but no known complete proof.
- Rating: < 150% conventional Over-Unity
- Leverage: 1:0.5 (Unconventional Reciprocal Wedges)
- Ball Mass = 1X (No counterweight)
- Maximum Gradient: < 22.5 degrees.

Escher Lever

- Date of First Evidence: May 12, 2017.
- State of affairs: Compelling evidence proper construction would create repeating cycles due to already proven automatic motion on four sides of a quadrilateral. Some creativity required.
- < 129% conventional Over-Unity.
- Leverage: : 3 - 4 : 1
- Counterweight Mass: 3X to 4X
- Maximum Gradient: < 13.05 degrees.

Escher Machine

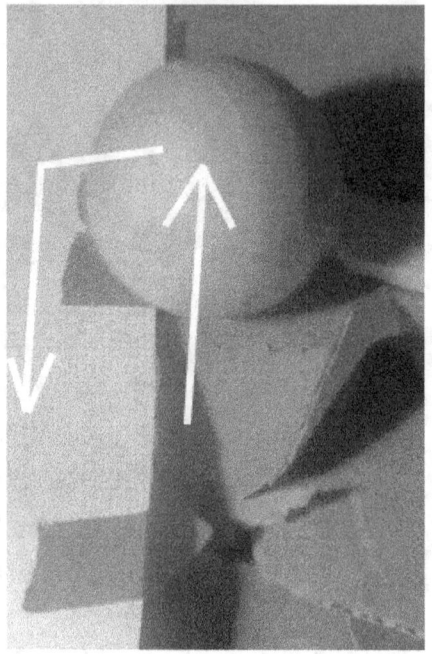

- Date of First Evidence: July 3, 2014.
- State of affairs: In most cases extremely difficult angles, and I mean extremely, beyond the average hobbyist. Despite doubts in 2021 the device has shown promising characteristics as shown. Bard AI says: "The device does seem to be able to produce over-unity, at least under certain conditions." And "[Referring to the Escher Machine] the object has a good chance of making contact with the backboard... which

44

would allow it to continue moving in a second cycle." --#BardAI

- Rating: < 150 % conventional Over-Unity
- Leverage = 1:0.5 (Unconventional Slope Acting on Wedge)
- Ball Mass = 1X (No counterweight)
- Maximum Gradient: Not extreme.
- Optimal Angle: 1.1025 percent degrees H X V +/- unknown [conjectured]

Exponential Balance

Automatic takeoff assistance, apparently over-unity.

ABOVE: From Nathan Coppedge's Simple Machines page.

Exponential Balance 3

ABOVE: An experiment with the Exponential Balance 3 showing movement of an equal object longer vertical distance using solely one input, from August 13, 2021.

Exponentially-Efficient Balance

Below: Exponentially Efficient Balance:

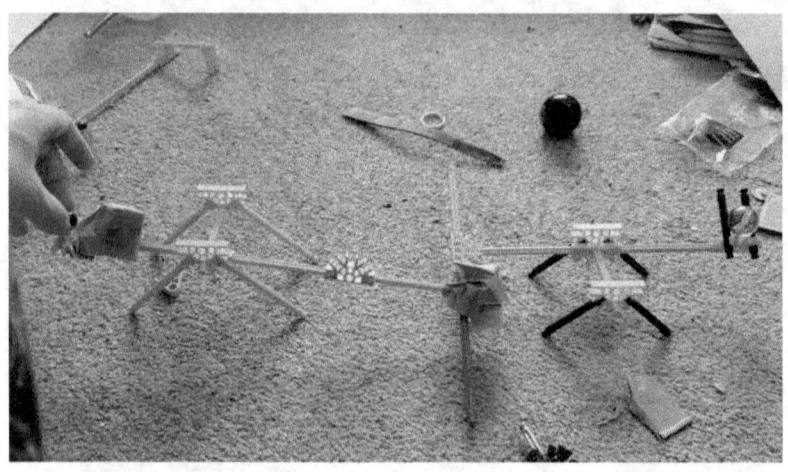

APPROXIMATION OF MATH:

Target mass: balance with 2 : 11 weight ratio

Trigger mass: Normally 1:3 leverage, 2:4 weight ratio

Foolishly Lucky Perpetual Motion

This design could be called the "Foolish" PMM
ATYPICAL PERPETUAL MOTION
Along the lines of the 2nd Equation

Nathan Coppedge   May 7, 2020

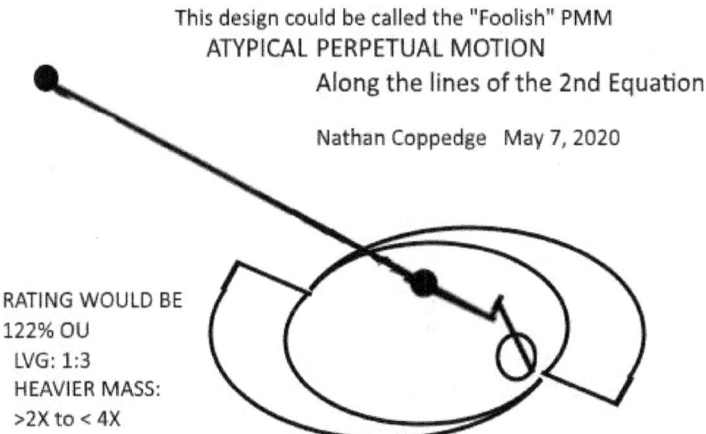

RATING WOULD BE
122% OU
 LVG: 1:3
 HEAVIER MASS:
 >2X to < 4X

THIS DESIGN THOUGHT NOT TO WORK FOR COMPLICATED REASONS

Hapt-Symmetric Vertical Lever

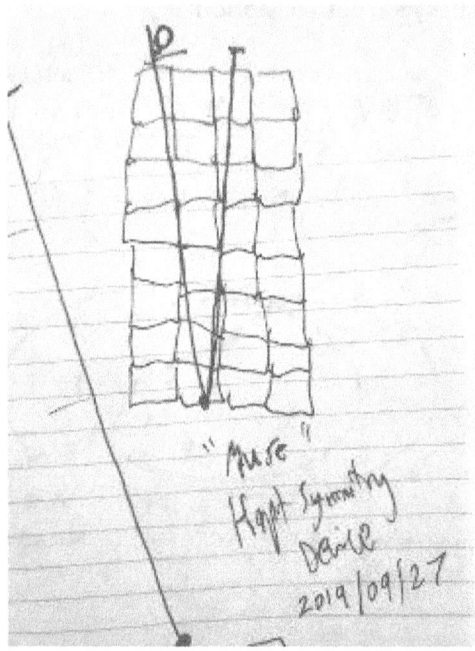

- Date of Invention: September 27, 2019.
- State of affairs: An intriguing idea, theoretically as functional as the Improved Vertical Lever, no known attempts or even similar models.
- Rating: < 150% conventional Over-Unity.
- Leverage: 1:1
- Counterweight Mass: >1.5X to <2X
- Maximum Gradient: < 22.5 degrees.

High Concept Machine

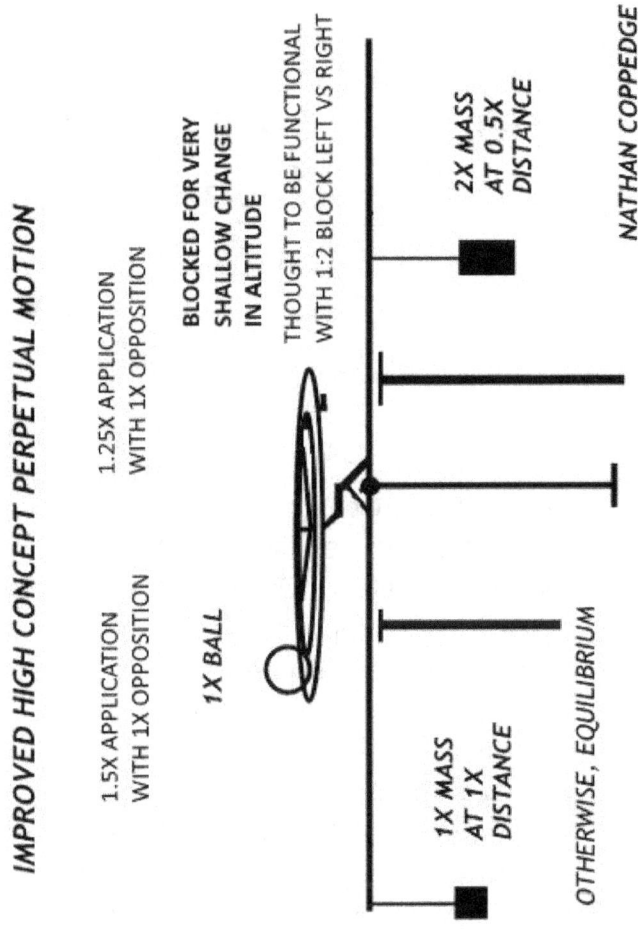

IMPROVED HIGH CONCEPT PERPETUAL MOTION

1.5X APPLICATION
WITH 1X OPPOSITION

1.25X APPLICATION
WITH 1X OPPOSITION

BLOCKED FOR VERY
SHALLOW CHANGE
IN ALTITUDE

THOUGHT TO BE FUNCTIONAL
WITH 1:2 BLOCK LEFT VS RIGHT

1X BALL

2X MASS
AT 0.5X
DISTANCE

1X MASS
AT 1X
DISTANCE

OTHERWISE, EQUILIBRIUM

NATHAN COPPEDGE

Hindu Device

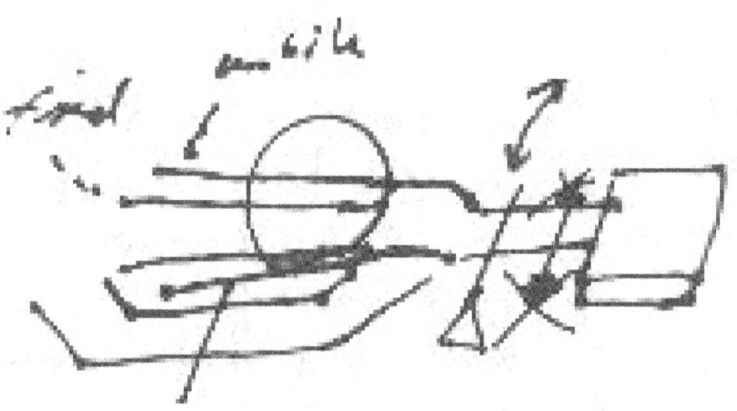

- Date of Invention: By August 14, 2019.
- Secretive clues: By December 29, 2016.
- State of affairs: One of the better uses of weight application for return, no known major experiments on this yet.
- Rating: < 120% conventional Over-Unity.
- Lvg: 1 to 1.5 : 1
- Counterweight Mass: >1.75X <2X
- Maximum Gradient: < 9 degrees.

Horizontal Swivel Wheel

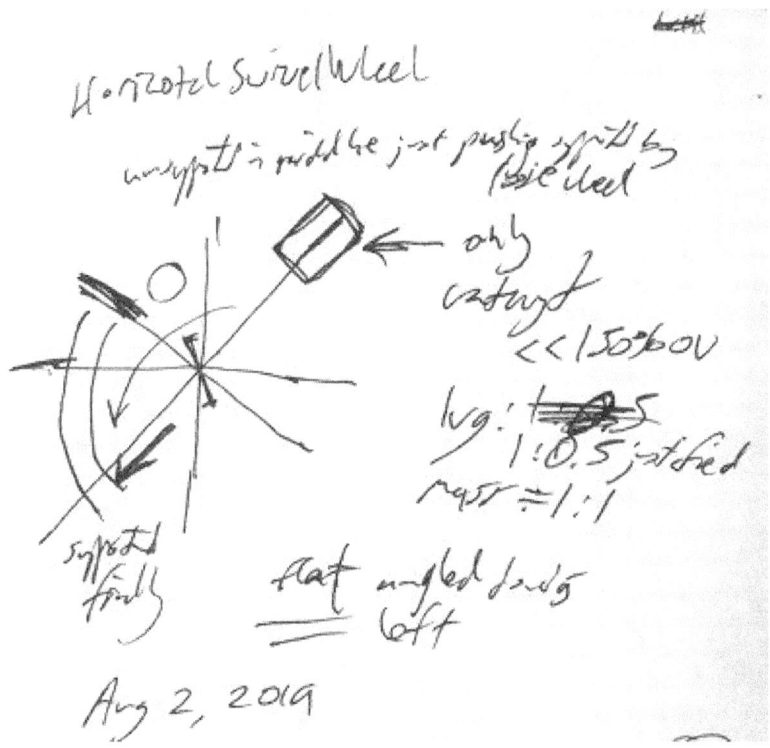

- Date of Invention: August 2, 2019.
- State of affairs: A clever design. Remember the lever-wheel is set at an angle that is almost horizontal, and that the lever wheel is constantly unbalanced in a particular way.
- Rating: < 150% conventional Over-Unity.
- Leverage: 1:0.5

- Counterweight Mass: (1:1)
- Maximum Gradient: < 22.5 degrees.

Inclementor Device

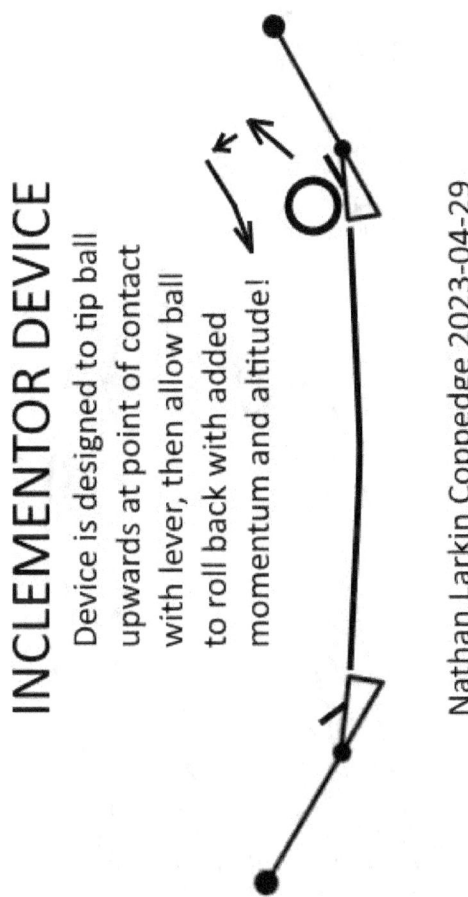

INCLEMENTOR DEVICE

Device is designed to tip ball upwards at point of contact with lever, then allow ball to roll back with added momentum and altitude!

Nathan Larkin Coppedge 2023-04-29

Jer Ram's Rouletto

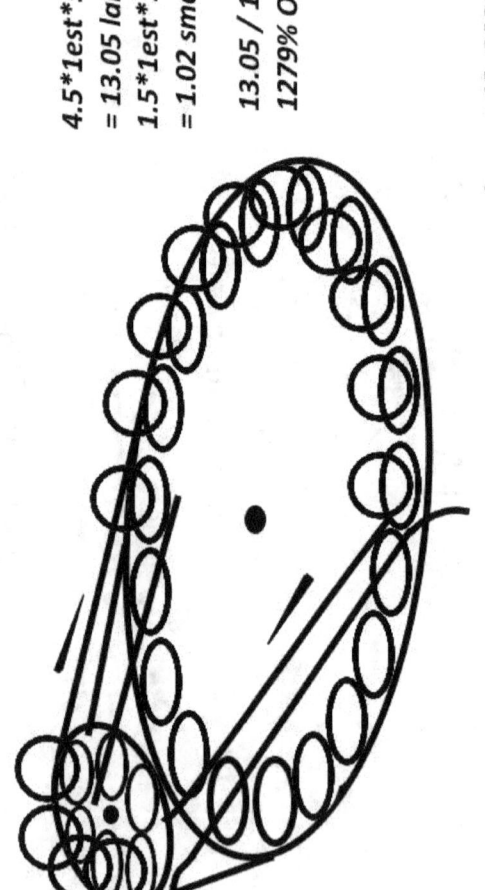

JER RAM'S ROULETTO
Diagram by Nathan Larkin Coppedge

4.5*1est*5*0.58
= 13.05 large wheel
1.5*1est*1*0.68
= 1.02 small wheel

13.05 / 1.02
1279% OU ???

August 19, 2022

Concept from 2009 - 2018 or earlier by Jer Ram

Jer Ram's Top-Heavy Conveyor

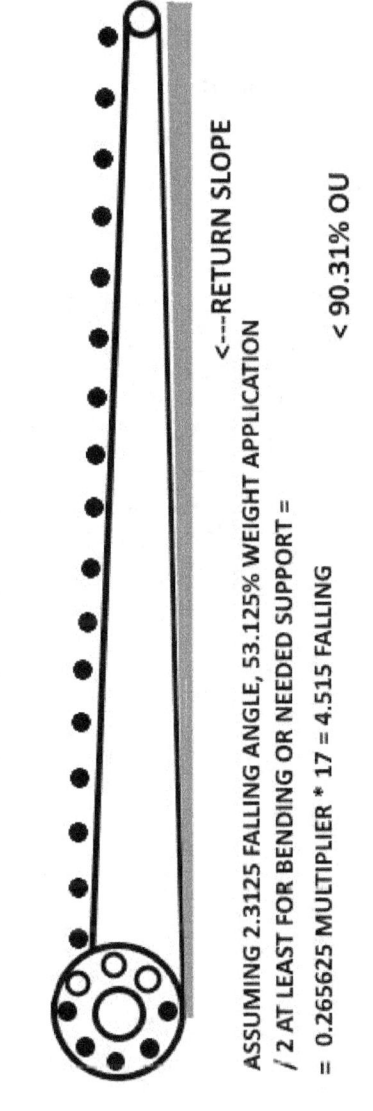

JER RAM'S TOP-HEAVY CONVEYOR
PERPETUAL MOTION MACHINE

<---RETURN SLOPE

ASSUMING 2.3125 FALLING ANGLE, 53.125% WEIGHT APPLICATION

/ 2 AT LEAST FOR BENDING OR NEEDED SUPPORT =

= 0.265625 MULTIPLIER * 17 = 4.515 FALLING

< 90.31% OU

5X MAX RISING BALLS = RATIO OF 4.515: 5 IN THIS CONFIGURATION

AGAINST OVER-UNITY

CONCEPT FROM 2004 - 2018 - MODIFIED 2022

J Roach Perpetual Motion

"Lazy Susan" Perpetual Motion

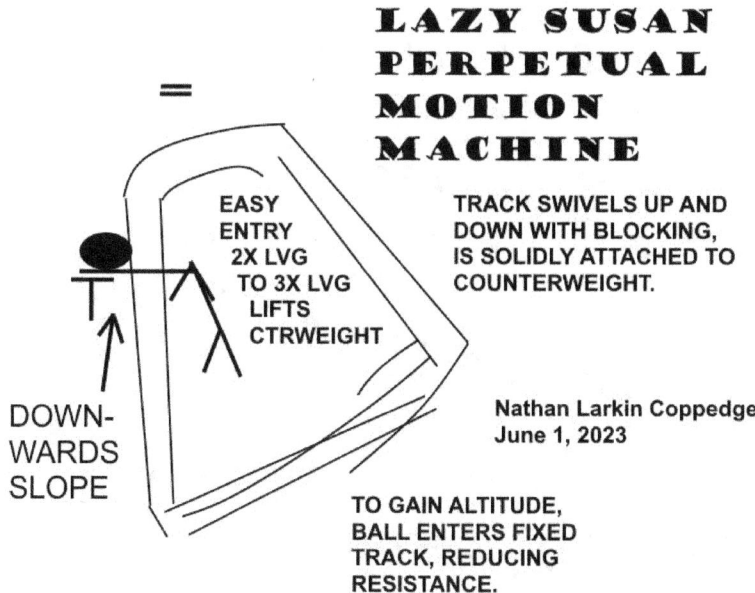

LAZY SUSAN PERPETUAL MOTION MACHINE

EASY
ENTRY
2X LVG
TO 3X LVG
LIFTS
CTRWEIGHT

TRACK SWIVELS UP AND
DOWN WITH BLOCKING,
IS SOLIDLY ATTACHED TO
COUNTERWEIGHT.

Nathan Larkin Coppedge
June 1, 2023

DOWN-
WARDS
SLOPE

TO GAIN ALTITUDE,
BALL ENTERS FIXED
TRACK, REDUCING
RESISTANCE.

Levers, Specially-Shaped, Hypothetical / Theoretical

MECHANICAL ARCHICULATES
of Proportive Bearing

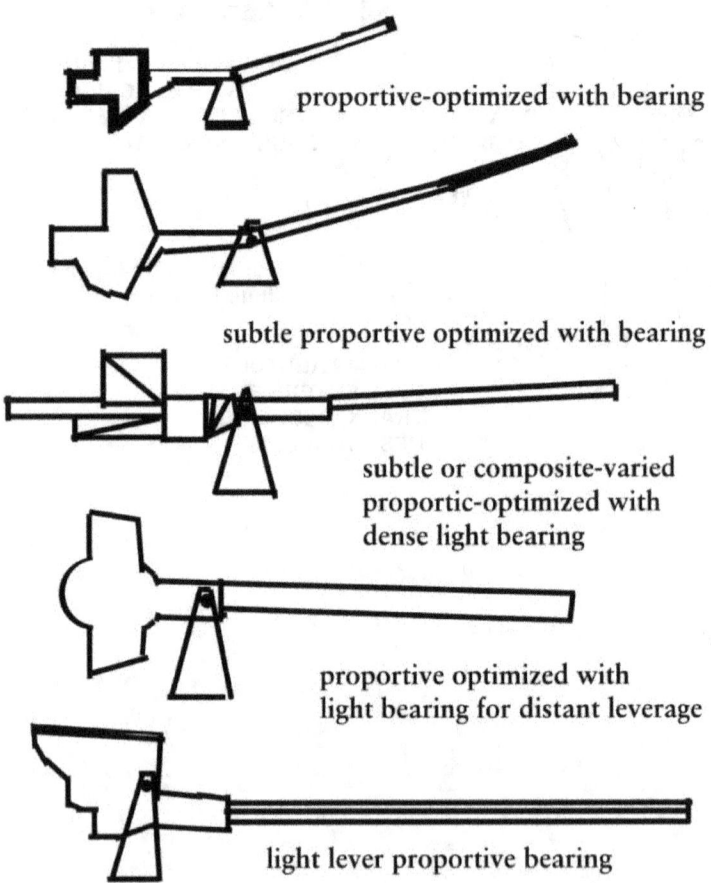

proportive-optimized with bearing

subtle proportive optimized with bearing

subtle or composite-varied
proportic-optimized with
dense light bearing

proportive optimized with
light bearing for distant leverage

light lever proportive bearing

Lighter Weight can Lift Heavier Weight

1/2 m x d experiment using lighter weight to lift heavier weight. In a balance an equal weight cannot create a continuous upward motion. Here a lighter weight creates a continuous upward motion of a heavier weight. It must be more efficient than a balance (Below). Let's say that's at least a little bit true, because we can imagine adding other properties which would make something with a similar or different type of efficiency more efficient.

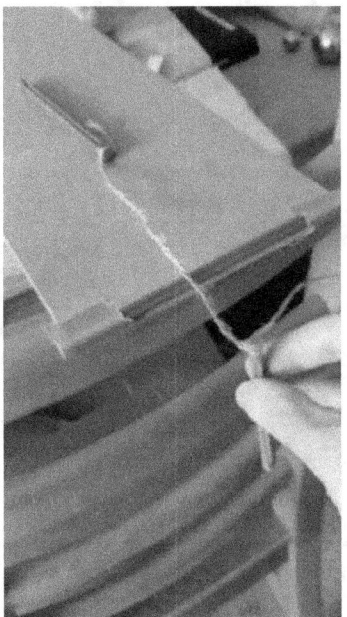

IMAGE: 1/2 Mass x Distance experiment (Coppedge, October 14, 2019).

Lucky Black Box

Magnet-Balance Combination

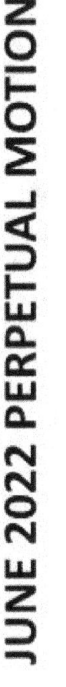
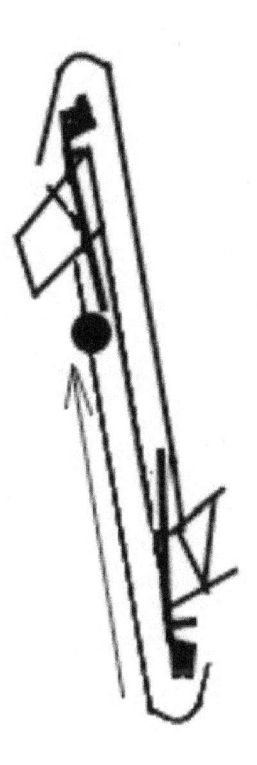

JUNE 2022 PERPETUAL MOTION

1x momentum X 6 = 6X momentum

Resistance = 6X - (3 * 0.9) = 2.7

6 - 2.7 = 3.3 energy

3.3 / 6 = +1 mass = 155% OU

155% - 16.6667% = 138.33% Conserv OU

2.7 : 6 resistance 155% OU

Magnet, Natural Magnet

**Natural Magnetic
Perpetual Motion
"Newton's Magnet"**

Remember to keep
the pull of the magnet
>0.65X the mass of
the ball

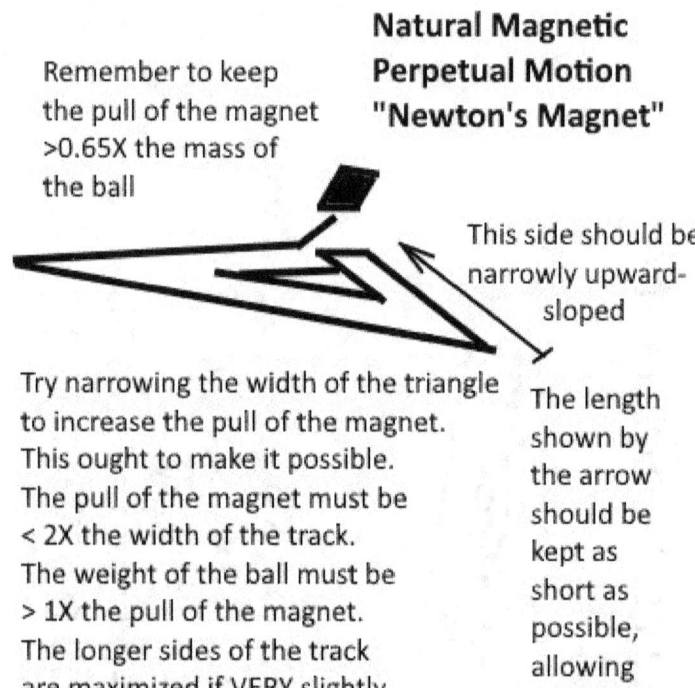

This side should be
narrowly upward-
sloped

Try narrowing the width of the triangle
to increase the pull of the magnet.
This ought to make it possible.
The pull of the magnet must be
< 2X the width of the track.
The weight of the ball must be
> 1X the pull of the magnet.
The longer sides of the track
are maximized if VERY slightly
downward-sloped when the
marble rolls counter-clockwise.

The length
shown by
the arrow
should be
kept as
short as
possible,
allowing
a weaker
magnet.

Nathan Larkin Coppedge

Magnetic Ramp

May 17, 2022 possible success with magnetic ramp: https://www.youtube.com/watch?v=WdBONGYq SEk

Marble Motor, Horizontal

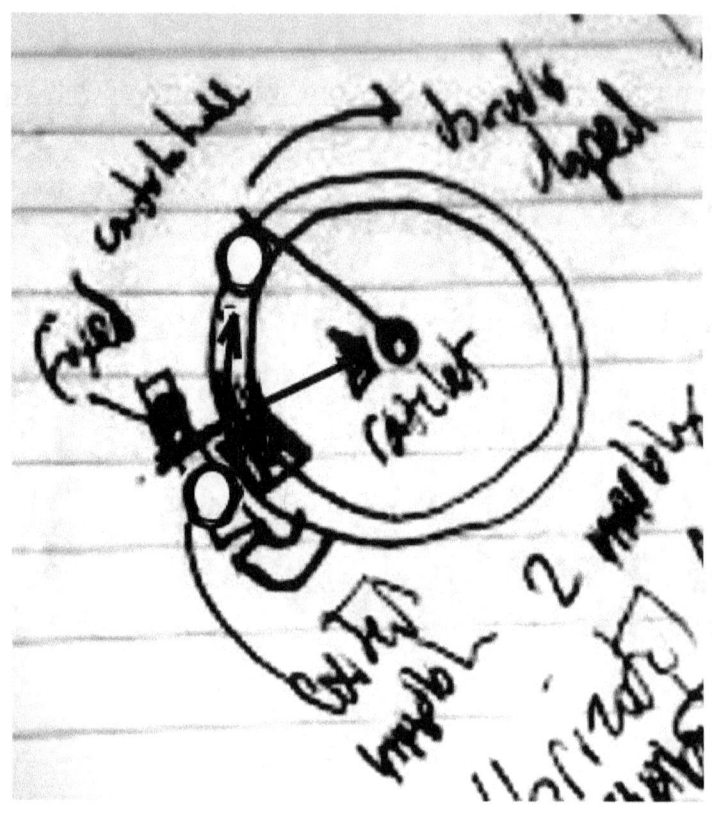

Marie Antoinette, Secret of

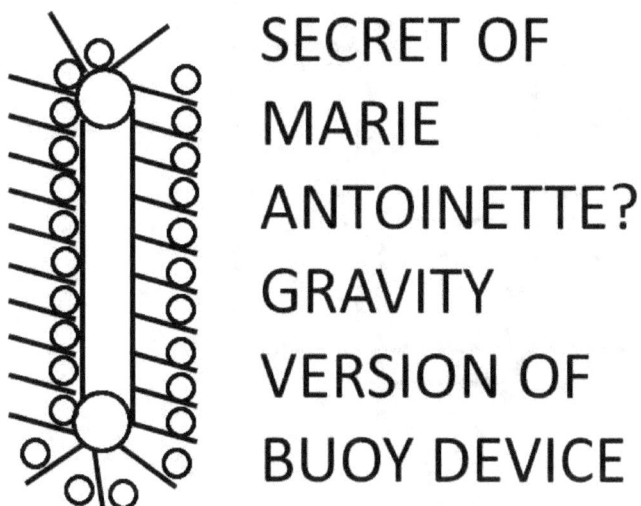

SECRET OF MARIE ANTOINETTE? GRAVITY VERSION OF BUOY DEVICE

Nathan Larkin Coppedge
June 7, 2021

Mathematical Perpetual Motion, 1st Generation

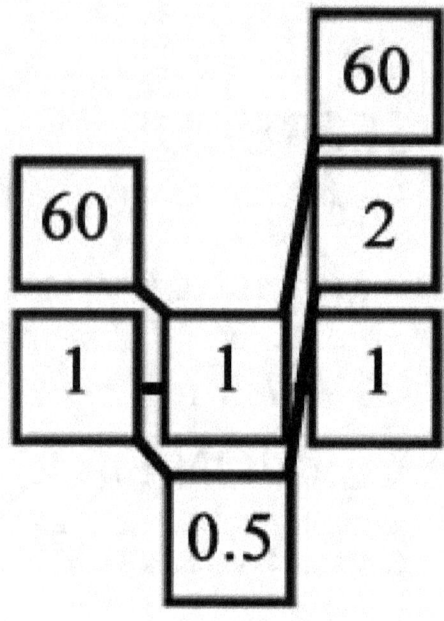

Mathematical Perpetual Motion, 2nd Generation

Volitional Energy = Mobile Units divided by Dual-Axial Units, Maximized when Dual-axial = Zero, which must be properly justified.

Volitional Equilibrium = Mod. Units divided by (Stems Per Cycle divided by Sub-Cycles Per Cycle), Maximized with a low number not usually zero.

Volitional Efficiency or Unconventional Over-Unity Rating = Volitional Energy Above divided by Volitional Equilibrium Above

Mathematical Perpetual Motion, 3rd Generation

EQUATIONS:
Min Heavier Mass = (Max Lvg / 2) + 1
Max Heavier Mass = Min Lvg + 1
Min Lvg = Max Heavier Mass - 1
Max Lvg = (Min Heavier Mass - 1) X 2
Over-Unity = Heavier Mass Rng /
      Lvg Ratio + 1 X 100 (%)
Smaller Mass = 1X
PM Cars Extra Mass < OU - 100%
Flying Vehicles Extra Mass < OU - 200%
Flying does not work when
  Lvg Rng >= 1/2 max leverage.
*Planetoids: Phi / 2 + 1 * 100*
    = < 130.9% Conventional Over-Unity

Medical Applications

Primary:

- Efficient Powders.
- Uneven balance.
- Movement with acceleration.
- Momentum with reduced resistance.
- Buoyancy displacement.
- Reciprocating application.

Applications of long-lived nanobots:

- Constant cell repair.
- Immunology.
- Brain function.
- Various body functions.

Larger-scale applications:

- Surgery.
- Power.
- Mobility.
- Enhanced implements.

Immortality Research

(1) A general logic extends the idea of exponential efficiency to higher and lower numbers of categories, an extension of the concept of total categories, as shown in the attachment. It is thought a variation on this involving 150% or 200% energy corresponding to human or post-human calorie use may be one of the most promising

variations in immortality research if applied correctly. It was observed before developing this logic that perhaps immortality is the use of the negative Function Spectrum for material purposes, whereas the magic is somehow the use of the positive Function Spectrum for abstract uses such as through high-dimensional implication of high authority.

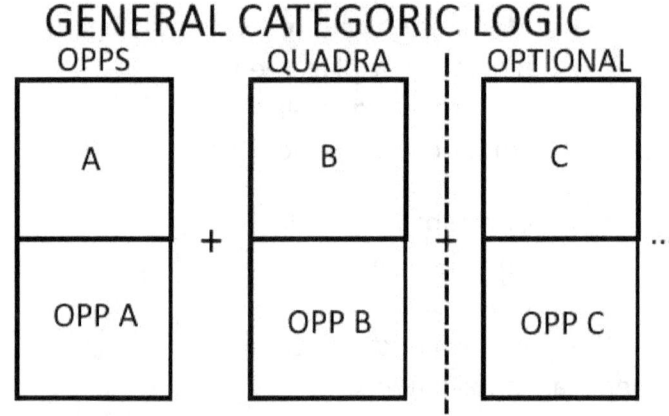

GENERAL CATEGORIC LOGIC

READ ACROSS, LEFT TO RIGHT, TOP TO BOTTOM

THIS SIMULATES THE CYCLICAL OPERATION

ANTITHEORY: READ ACROSS, RIGHT TO LEFT, TOP TO BOTTOM (B, A... OPP B, OPP A...)

Example with Two Categories and 0 Diff, 0 Eff, 0 Results:
Using Diff Equation (Opposites): Function Spectrum

Example with Four Categories and 3 Eff, -1 Diff, 2 Results
Using TOE Equation (Variation): ABCD and ADCB
Example with Six Categories and 2 Results -1 Diff, 3 Eff:
Using Immortality (2-Variation):
Assuming problem if Death Dies it's Mortal
Assuming solution, Living Eternally is Immortal
Assuming a problem, If trying to try is trying
Assuming a solution if perishing of perishing perishes

(2) It was estimated that Earth Years, a description of human health ratings, could be estimated using energy (1.5) * 60 to predict 90 years of life for typical long-lived humans. This equation was adapted to the perpetual motion equations in this file: Universal Metaphysical Medicine: UniversalMetaphysicalMedicine.xlsx where it was found inputting categories close to one has the highest energy and consequently the highest health, of ideally close to infinite, whereas this behavior is exhibited no where else in the same diagram. In fact, the abstract efficiency modifier statistic which is different from the health rating, is predicted to sink to approximately negative infinity / 2 at the same increment of categories. The ratings of course are quite optimistic as no successful applications are known, and research on the master angle is still in its initial stages. Encouragingly, there is a further suggestion that a coherent or hyper-efficient approach is somehow conducive to immortality objectively, and that coherence expresses the highest efficiency possible in reality. This also suggests that high efficiency, and specifically high coherent efficiency, is correlated with immortality.

(3) After working on the Metaphysical Medicine document, I concluded that immortality was correlated with coherence in the 2nd dimension or

in fact, the perfection of any given dimension, because one category means coherence, and all the infinite longevity values appeared to fall under one category. However, I had also found that in the Efficiency Formula which was thought to be an approximation of the immortality formula, the optimal efficiency was found to be 2 in the System Investigator Widget, while the immortal dimensions was found to be -2. The Dimensional ToE predicted that knowledge and perpetual motion were unified in the 3rd dimension, and that the 3rd dimension had an efficiency of 3 when the 3rd was correlated with the ToE or perpetual motion formula. It was suggested reductively the 1st and -1st were unified in what was called Efficiency 2 which was correlated with immortality, in the view that the 2nd was really the -1st and was unified with the 3rd, and that the -2nd was correlated with immortality. This lined up with the idea that Linguistics was correlated with immortality and also that coherence was centered around zero, and exhibited an infinite upward curve as often happens with zero in a balanced graph as would be possible with coherence. If 2 was the efficiency in the -2nd dimension of immortality, and the efficiency for the 3rd dimension was known to be 3, this suggested the -3rd dimension of immortality was correlated with the 3rd dimension, and so on up and down the Function Spectrum (-4th dimension of immortality could be correlated with the 1st dimension, under the observation that -4 had to do with 'leaving the system' and the 1st dimension had been found to be related to death and evolution). This meant zero was truly zero (ignored in the Function Spectrum), and that dimensions were likely the things that were immortal. It also suggested a reciprocal relationship between Efficiency and Immortal Dimensions which indicated that difference >= 100% is inspired by exponential efficiency (related to negative differences and higher positive efficiency, with the efficiency being optimally higher by a rate of 3:1), and immortality now correlated with efficiency >= 100% (?) may eventually be inspired by mechanical efficiency (related to positive differences and possibly positive efficiency, with the efficiency being optimally higher by a rate of 2:1).

Perfection of the dimension + desired attribute = immortal same attribute in same dimension.

Coherence + health = Immortal health in the 2nd dimension.

Perpetual motion + health = Immortal health in the 3rd dimension

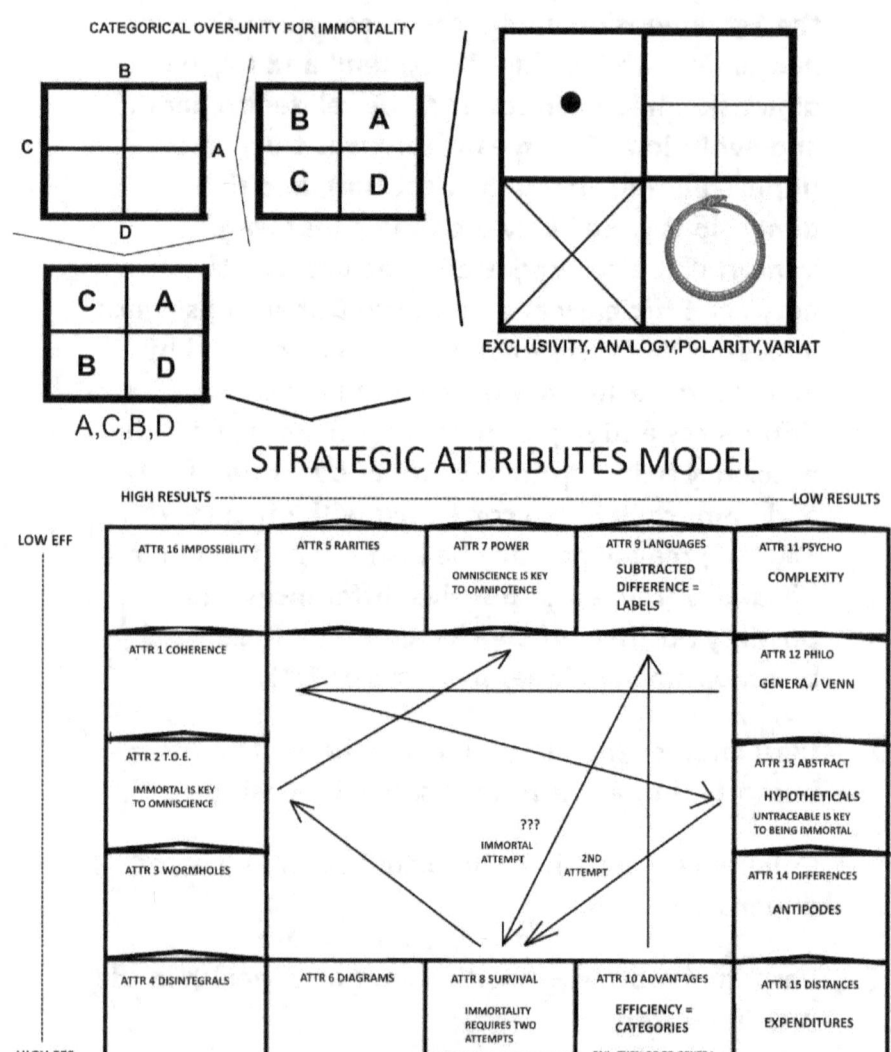

CATEGORICAL OVER-UNITY FOR IMMORTALITY

A,C,B,D

EXCLUSIVITY, ANALOGY,POLARITY,VARIAT

STRATEGIC ATTRIBUTES MODEL

HIGH RESULTS ---LOW RESULTS

ATTR 16 IMPOSSIBILITY	ATTR 5 RARITIES	ATTR 7 POWER OMNISCIENCE IS KEY TO OMNIPOTENCE	ATTR 9 LANGUAGES SUBTRACTED DIFFERENCE = LABELS	ATTR 11 PSYCHO COMPLEXITY
ATTR 1 COHERENCE				ATTR 12 PHILO GENERA / VENN
ATTR 2 T.O.E. IMMORTAL IS KEY TO OMNISCIENCE		??? IMMORTAL ATTEMPT	2ND ATTEMPT	ATTR 13 ABSTRACT HYPOTHETICALS UNTRACEABLE IS KEY TO BEING IMMORTAL
ATTR 3 WORMHOLES				ATTR 14 DIFFERENCES ANTIPODES
ATTR 4 DISINTEGRALS	ATTR 6 DIAGRAMS	ATTR 8 SURVIVAL IMMORTALITY REQUIRES TWO ATTEMPTS 1ST ATTEMPT: GO TO GENERA	ATTR 10 ADVANTAGES EFFICIENCY = CATEGORIES FAIL: THEN GO TO GENERA	ATTR 15 DISTANCES EXPENDITURES

LOW EFF

HIGH EFF

IMPOSSIBLE IS POSSIBLE, DEATH, LIFE: THE TRUE AMBERGRIS

Mini Perpetual Motion

(March 18, 2018):

I see some developments in this area in the next 180 - 220 years, or perhaps more accurately the area of large nanobots, dynamic techno-textures, or miniature machines (between nano and meso).

Applications might include things like:

- Surfaces that generate energy like solar panels without requiring sunlight.
- Systems that never go to zero power.
- Nanobots that carry internal energy without a battery or conventional power source.
- Ubiquitous self-powered digital surfaces initiating the age of physical environments that react to thought.

Modular Lever, Improved

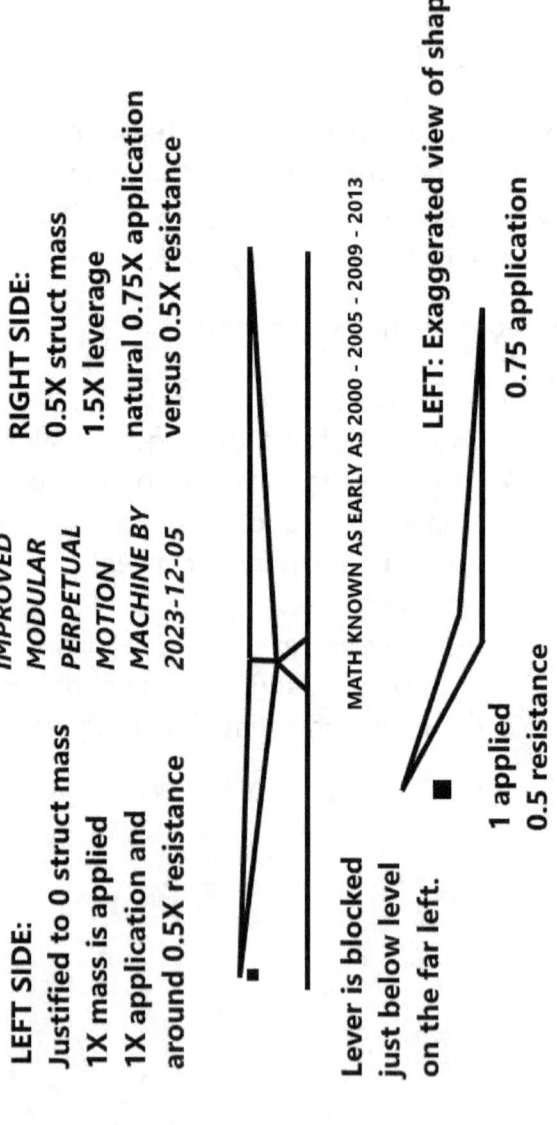

LEFT SIDE:
Justified to 0 struct mass
1X mass is applied
1X application and
around 0.5X resistance

IMPROVED MODULAR PERPETUAL MOTION MACHINE BY 2023-12-05

RIGHT SIDE:
0.5X struct mass
1.5X leverage
natural 0.75X application
versus 0.5X resistance

LEFT: Exaggerated view of shape.

MATH KNOWN AS EARLY AS 2000 - 2005 - 2009 - 2013

0.75 application

1 applied
0.5 resistance

Lever is blocked
just below level
on the far left.

Modular Magnet Device, Russian

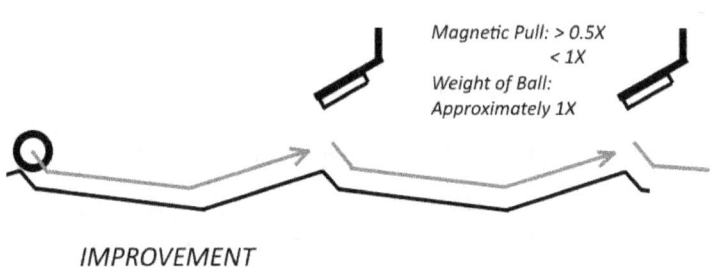

Magnetic Pull: > 0.5X
< 1X
Weight of Ball:
Approximately 1X

IMPROVEMENT OF THE RUSSIAN MAGNET DEVICE

Nathan Coppedge
2024-02-21

Modular Mass Machine (MMM),

See Dual Seesaw Device. Multistory Transportation Device

MULTISTORY TRANSPORTATION DEVICE

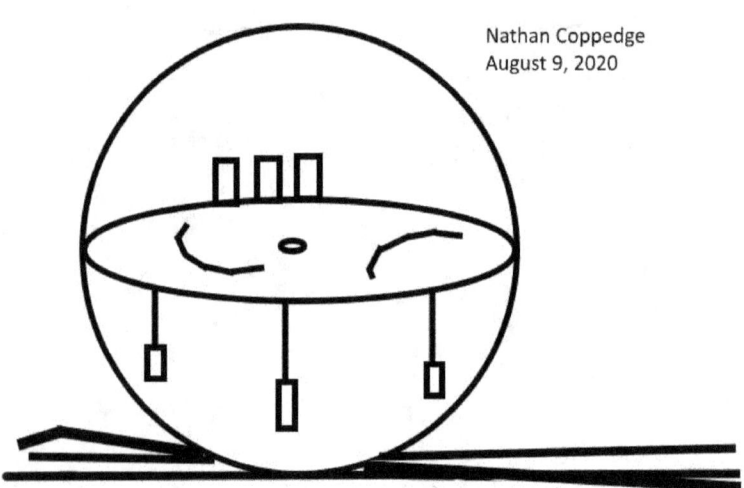

Nathan Coppedge
August 9, 2020

Nano Perpetual Motion

Until recently... this discipline did not really exist.

Options:

1. Some type of polarized hole.

2. Sites on larger machines.

3. Combinations of material textures such as magnet hairs and waves of relatively heavier parts.

—3rd Revised Periodic Table of Working Perpetual Motion Machines

Natural Torque 360 Degree Motion Modification

For the moment depends on movement of the base.

DETAIL OF TORQUE 360 DEGREES CAPABILITY

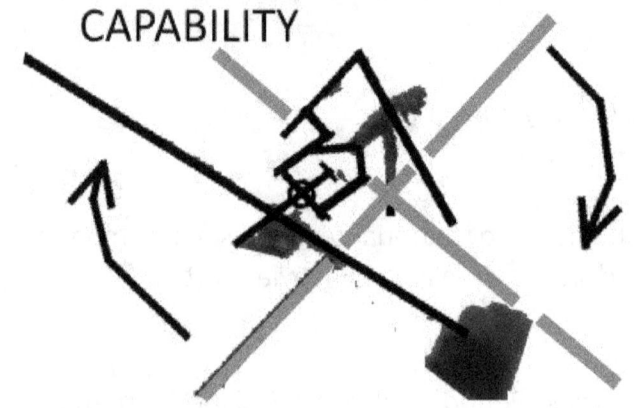

EXPERIMENT 2021-12-11   NCOPPEDGE

Natural Torque Device

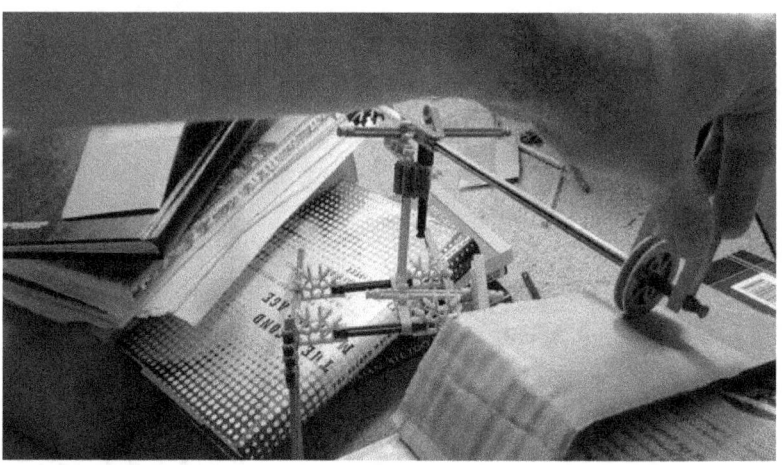

NIBW 1 & 5

**MODULAR N.I.B.W. 5   A. STEEP, B. SHALLOW,
MODULAR DESIGNED FOR FULL RECOVERY OF ALTITUDE**

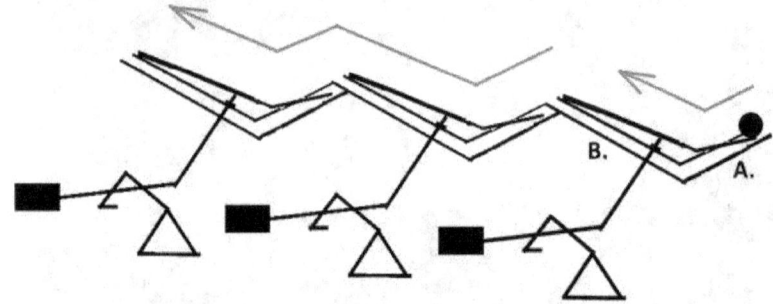

*ACTIVATED ENTIRELY BY SLOPE OF BALL VERSUS COUNTERWEIGHT
WITH SUPPORT FOR BALL AT EVERY POINT CREATING EFFICIENCY*

- Date of Invention: January 1, 2016.
- State of affairs: A bit promising but experimental setbacks. May require shallower angles.
- Rating: < 120% conventional Over-Unity.
- Leverage: 1 to 1.5 : 1
- Counterweight Mass: > 1.75X < 2X
- Maximum Gradient: < 9 degrees.

NIBW2, EVOLVED FORM

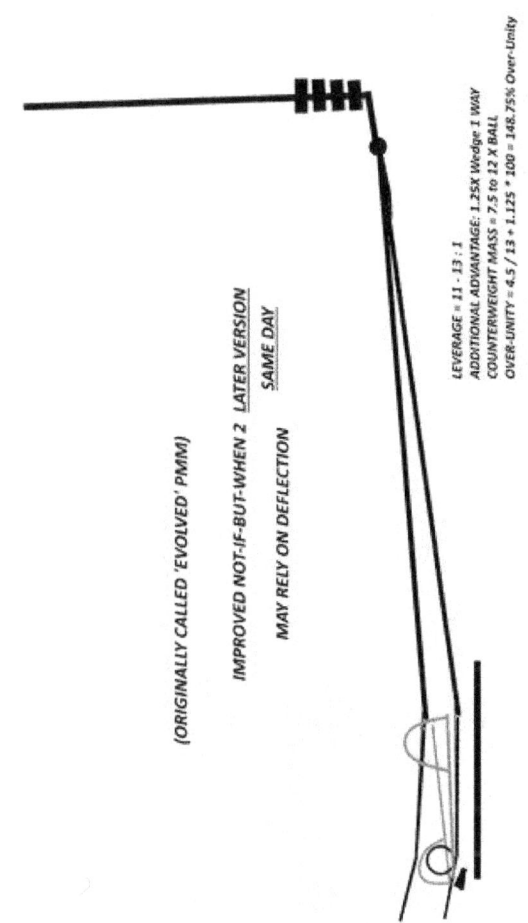

(ORIGINALLY CALLED 'EVOLVED' PMM)

IMPROVED NOT-IF-BUT-WHEN 2 LATER VERSION

MAY RELY ON DEFLECTION   SAME DAY

LEVERAGE = 11 - 13 : 1
ADDITIONAL ADVANTAGE: 1.25X Wedge 1 WAY
COUNTERWEIGHT MASS = 7.5 to 12 X BALL
OVER-UNITY = 4.5 / 13 + 1.125 * 100 = 148.75% Over-Unity

Nathan Larkin Coppedge 2022-04-04
Some prior work by Jorge Vargas

Leverage: 11 - 13: 1

Additional advantage: 1.25X wedge one way.

Counterweight: 7.5 to 12X ball.

Over-Unity: 4.5 / 13 + 1.125 * 100 = < 148.75%

Not If But When Machine Type 4

"NOT-IF-BUT-WHEN" MACHINE #4

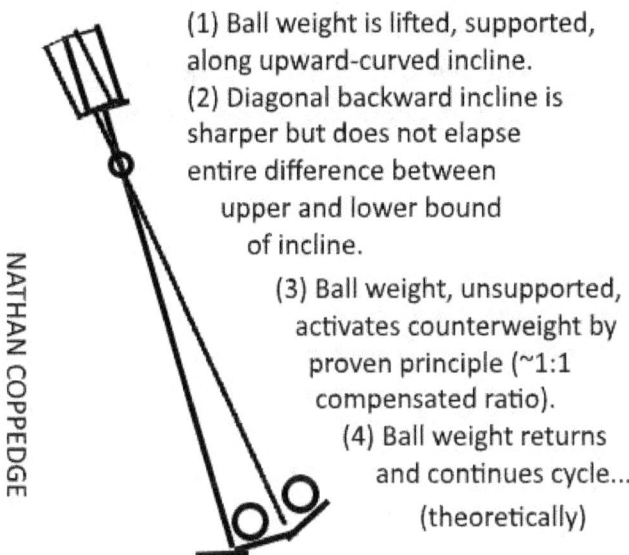

(1) Ball weight is lifted, supported, along upward-curved incline.
(2) Diagonal backward incline is sharper but does not elapse entire difference between upper and lower bound of incline.

(3) Ball weight, unsupported, activates counterweight by proven principle (~1:1 compensated ratio).
(4) Ball weight returns and continues cycle...
(theoretically)

NATHAN COPPEDGE

- Clarification of diagram (2020/06/19 or much earlier): It is now thought deflection alone plus return in an identical arc could be sufficient.
- Probable Date of Invention: January 10, 2016.
- Date of First Evidence: March 18, 2016.

- State of affairs: Some strong partial evidence.
- Rating: < 150% conventional Over-Unity
- Leverage = 3:1
- Counterweight Mass = >2.5X to <4X
- Maximum Gradient: < 22.5 degrees.

NIBW6

NOT-IF-BUT-WHEN MACHINE #6

NATHAN
COPPEDGE

C. B.
A.

WEIGHT RATIO 1:1 COMPENSATED
LEVER RATIO 1: 2.4 - 2.7+

AT PT. A COUNTER-WEIGHTED
CIRCULAR WIRE (LIGHTER COLOR)
APPLIES ANGLED PRESSURE
AGAINST BALL WEIGHT. BALL WEIGHT
RISES BECAUSE IT IS SUPPORTED.

PROCESS CONTINUES UNTL BALL REACHES
APEX B, AT WHICH POINT CIRCULAR WIRE
ANGLE CHANGES TO A STEEP UPWARD
INCLINE. HOWEVER, BALL APPLIES SIGNIFICANT
PRESSURE ON WIRE BECAUSE IT IS NOW ALMOST
FULLY UNSUPPORTED DUE TO STEEP ANGLE
C. SINCE BALL AND COUNTERWEIGHT ARE
APPROXIMATELY EQUAL IN COMPENSATED MASS
BALL WEIGHT SINCE IT HAS GREATER LEVERAGE
CAN CONTINUE PROCESS.

- Date of Invention: March 31, 2016.
- Date of First Evidence: July 6, 2018.
- Rating: Adjusted for larger mass < 119.23% Conventional OU, formerly rated at < 162.5% in terms of 1X mass (It is thought the key here is a highly vertical spiral dropped by full non-support).
- Leverage: 1 to 2 : 3.
- Ball Mass: >2.5X <4X
- Maximum Gradient: Approx < 9 degrees (not calculated).

Over-Unity Cars, 1ˢᵗ Generation

- Cheaper transportation (free energy vehicles!).

"BIG RIG" COASTER BICYCLE

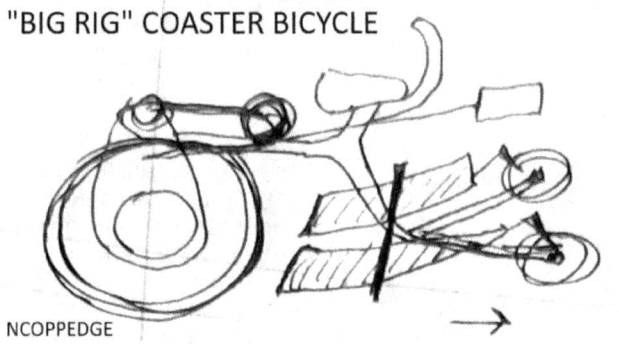

NCOPPEDGE

NCOPPEDGE

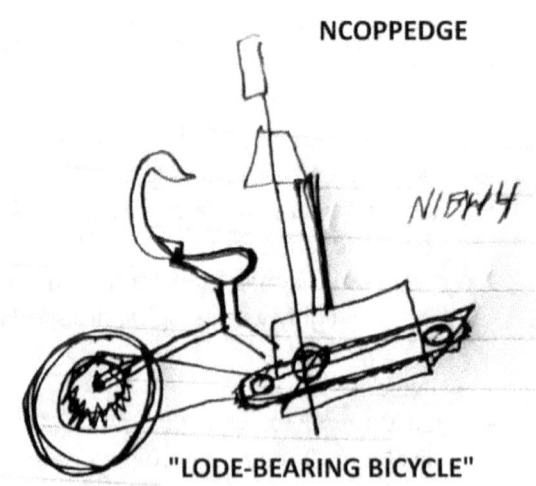

"LODE-BEARING BICYCLE"

Over-Unity Car-Wagon

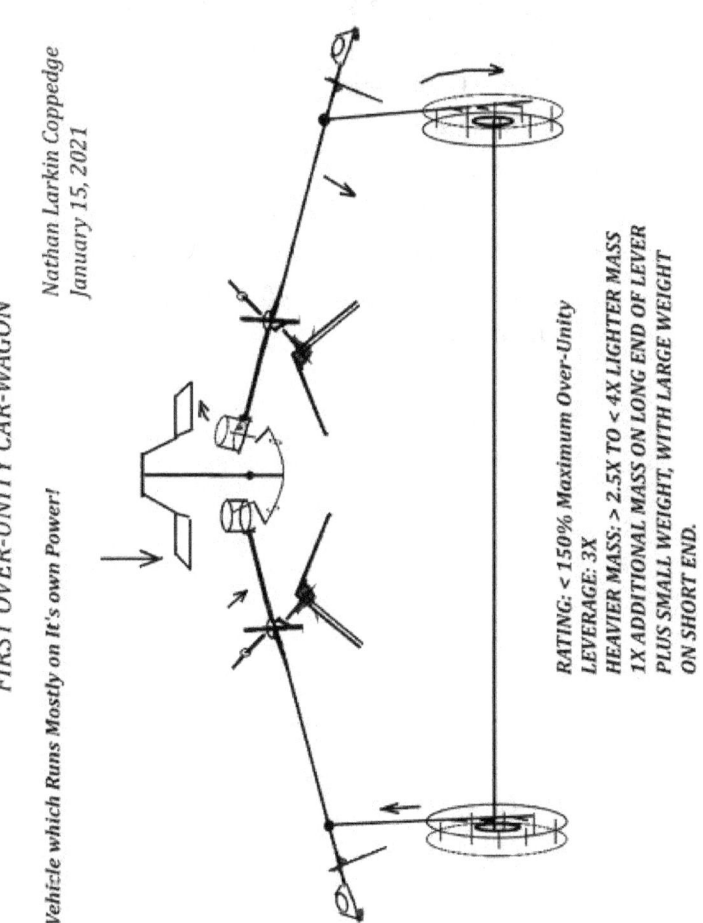

FIRST OVER-UNITY CAR-WAGON

Nathan Larkin Coppedge
January 15, 2021

A Vehicle which Runs Mostly on It's own Power!

RATING: < 150% Maximum Over-Unity
LEVERAGE: 3X
HEAVIER MASS: > 2.5X TO < 4X LIGHTER MASS
1X ADDITIONAL MASS ON LONG END OF LEVER
PLUS SMALL WEIGHT, WITH LARGE WEIGHT
ON SHORT END.

- Note: This device is not currently self-powered, instead it uses minimal input from the pedals using a property known as the difference drive.
- Over-Unity Rating: < 150% Conventional OU
- Leverage: 3:1
- Counterweight: > 2.5X to < 4X
- Relevant Formulas: Mostly standard, except the mechanism is not yet self-powered.

Over-Unity Car Wagon 2: Efficiency Car

EFFICIENCY CAR (Car-Wagon 2)

Counter-Weight causes motion of wheels,
after leverage is applied!

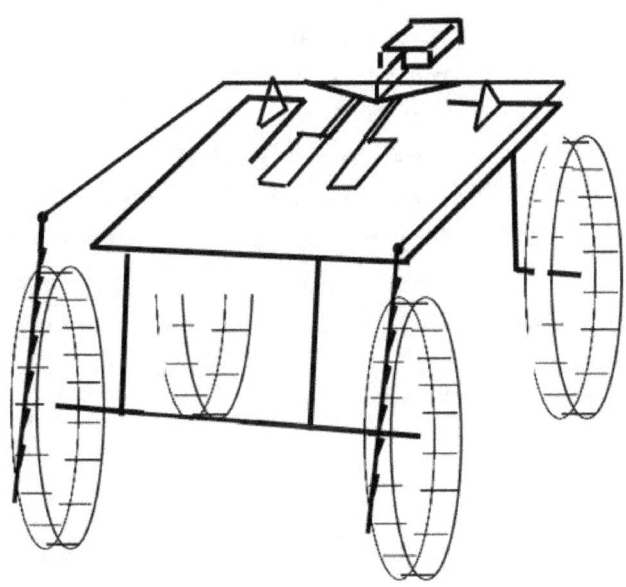

NATHAN COPPEDGE JANUARY 18, 2021

- Note: This device is not currently self-powered, instead it uses minimal input from the pedals using exponential efficiency.
- Over-Unity Rating: < 150% Conventional OU
- Leverage: 2:1
- Counterweight: > 2X to < 3X
- Relevant Formulas: Mostly standard, except the mechanism is not yet self-powered.

Parallaxial Slope Device

PARALLAXIAL SLOPE PERPETUAL MOTION MACHINE

Slope A. is less than Slope B. (both constant),
yet A. permits circuit;

Difference of C. can
be overcome by
momentum, because
B. exceeds A.

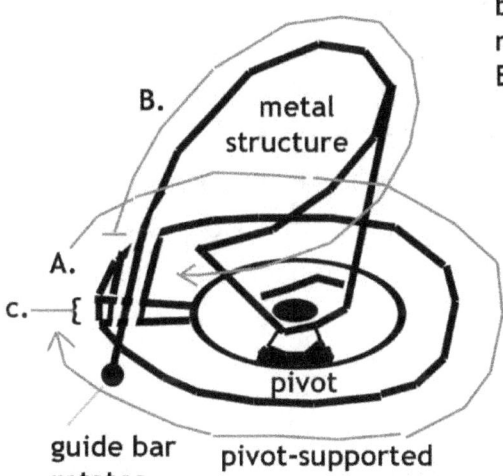

An angular slot at C.
is designed to aid
recovery from slope
while allowing a return
to the superior angle.

A continuous
fixed guide slot
may be added
around the
structure as
necessary.

B. metal structure

A.

c.

pivot

guide bar rotates 1/360

pivot-supported during second leg

N. Coppedge

Permanent Energy Concept

More efficient than perpetual motion. Properties: Undepletable source relative to a variety of time-frames. Term should not be abused, must be defined as more advanced than component chain-reaction chemical perpetual motion. Sense of timelessness, eternity, non-depletability. Possible positive effects like immortality, special usefulness. Likely based on perpetual motion nanostructures. Not a worthwhile investment at first, like rockets compared to siege engines.

Perpetual Motion Appliances

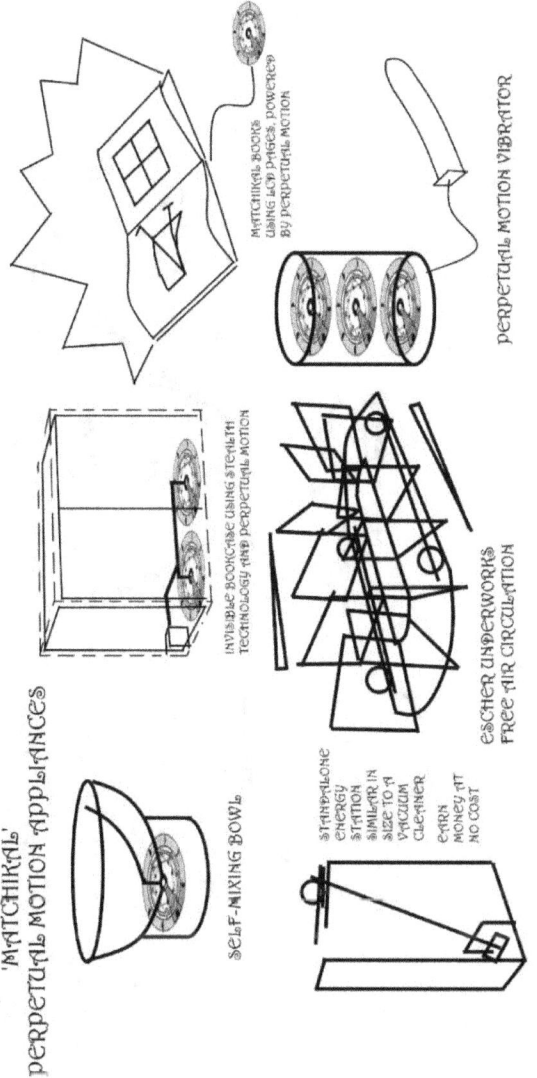

MATCHICAL BOOKS
USING LED PAGES, POWERED
BY PERPETUAL MOTION

PERPETUAL MOTION VIBRATOR

INVISIBLE BOOKCASE USING STEALTH
TECHNOLOGY AND PERPETUAL MOTION

ESCHER UNDERWORKS
FREE AIR CIRCULATION

'MATCHICAL'
PERPETUAL MOTION APPLIANCES

SELF-MIXING BOWL

STANDALONE
ENERGY
STATION
SIMILAR IN
SIZE TO A
VACUUM
CLEANER

EARN
MONEY AT
NO COST

Perpetual Motion Bicycle, 1st Generation

"BREAK" THE CYCLE

INVENTORBIKE

NATHAN COPPEDGE
July 4, 2021

Perpetual Motion Bow Machine 2

BOW PERPETUAL MOTION 2

BOWSTRING ROTATES USING SPIRAL TO LIFT BALL.

BALL IS ACTUALLY FIRMLY ATTACHED TO BOWSTRING AT CLOSE DISTANCE

BALL WEIGHS 1X

BOWSTRING IS >0.5X <1X

TRACK HAS VERY SHALLOW GRADIENT

IDEAL OVER-UNITY WITHOUT GRADIENT = < 150%

REMAINING PUZZLE: HOW TO FIX ISLAND

PERHAPS USE BAR WITHIN CENTRAL SPINDLE THAT EMERGES IN THE CENTER, WITH SEPARATE ROTATION ON TOP AND BOTTOM, OR HAVE STRING ROTATE AT TOP AND BOTTOM BUT NOT IN THE MIDDLE AT ALL.

Nathan Larkin Coppedge
March 11, 2023

Perpetual Motion Crossbow, 1ˢᵗ Generation

Panjagān

Probably meaning
'knobby javellin thrower'
was an early repeating weapon
theorized to use one string
and a series of rising arrow
compartments to send a series of
four or five or more arrows
either quickly or slowly
or anywhere between.
The string usually rests
on a series of steps preceding each
arrow which can be assisted by a
very slightly backward-slanted knob
before each step which prevents
misfiring.

Perpetual Motion Defense Weapons

OVER-UNITY WEAPONS

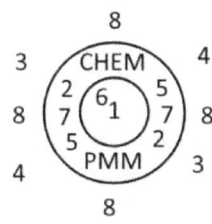

CHEM PMM MAY ALSO DEPEND
ON THIS SORT OF REACTION
BEFORE EXISTING ITSELF
OR IT MAY NOT.

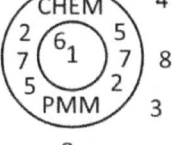

PERPETUAL MOTION BOMB "ETERNAL FIRE"

CRITICAL IS THE PRIOR
INVENTION OF CHEM PMM.
OTHERWISE, THE INVENTION
ACTS THROUGH PURE
CHEMISTRY EXCEPT
WITH AN ENDURING CORE,
AND REPEATED EFFECT
WHICH DEPENDS ON
EXTERNAL HEAT ACTING
ON THE OUTER CORE
TO REIGNITE THE CENTER,
THEN CAUSING A MORE
EXTREME REACTION WITH
CHEM PMM WHICH IS THOUGHT
TO BE SUSTAINABLE ONCE
CHEM PMM EXISTS.

SELF-RECHARGING LASERS

ONE OF THE FIRST 'IMMORTAL
DEFENSES' EVER CONCEIVED,
THE RELATIVE PERMANENCE OF
A WELL-BUILT LASER WHEN
COMBINED WITH AN PERMANENT
ENERGY SOURCE YIELDS LASERS
WHICH REQUIRE LITTLE TO NO
UPKEEP AND WHICH PERIODICALLY
RECHARGE THEIR ENERGY, OR CAN
STORE LARGE CAPACITIES THROUGH
DOWNTIME.

ESCHER RIFLING

AT LEFT IS AN APPROXIMATION
OF THE 1.1025 PERCENT DEGREES
H X V NECESSARY TO CREATE
AN OVER-UNITY GUN USING
CONVENTIONAL PROPELLANT.

IMPROVED DESIGN

KINETIC RECOILLESS RIFLE

USE OF AN ANGULAR CONNECTION TO
THE BACK OF BARREL COMBINED WITH
FORWARD LEVERAGE ON PROPELLANT MAY
CREATE DOWNWARDS RECOIL OR NO RECOIL
AT ALL THROUGH USE OF A PRINCIPLE
SIMILAR TO THE EXP GRAVITY LEVER.

ENERGY ARMIES

AN ARMY WITH AN UNLIMITED ENERGY BUDGET USING
PERPETUAL MOTION COULD POTENTIALLY BE OVERWHELMING.

"FREE-WEAPONS"

INCREASING A MILITARY'S ENERGY BUDGET MAY CREATE RESULTS
SUCH AS INCREASED CONVEYANCE, INCREASED WEAPONS
PERFORMANCE, QUICKER ENGAGEMENT, AND IMPROVED
ECONOMICS OF WARFARE.

Perpetual Motion Gravity Hammer (Exp Gravity Hammer)

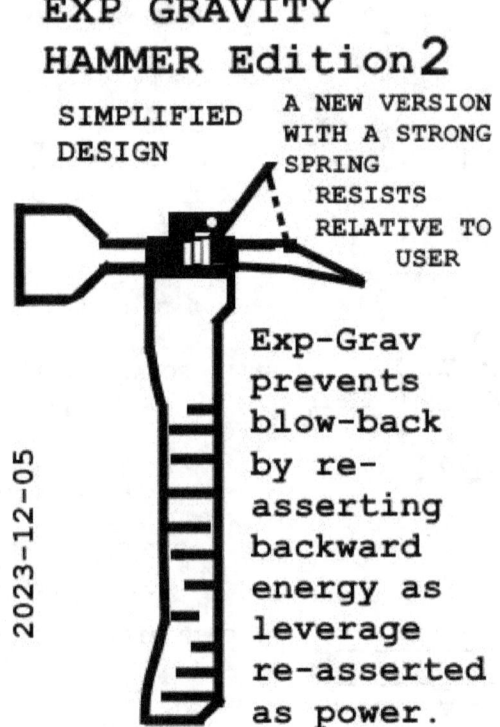

EXP GRAVITY
HAMMER Edition 2

SIMPLIFIED
DESIGN

A NEW VERSION
WITH A STRONG
SPRING
RESISTS
RELATIVE TO
USER

2023-12-05

Exp-Grav
prevents
blow-back
by re-
asserting
backward
energy as
leverage
re-asserted
as power.

Perpetual Motion Magical Staff of Lightning

PERPETUAL MOTION MAGICAL STAFF OF LIGHTNING

When set upright, this staff magically 'recharges' using a medium-size small perpetual motion machine called a Tilt Motor, designed by Nathan Larkin Coppedge.

The staff is designed as a pseudo-scientific self-defense staff which is filled with felt-like material and a long, winding copper filament.

This is a legitimate self-defense mechanism as once charged automatically, it may release significant amounts of electricity upon contact. It must be used carefully.

The Staff Concept was originally conceived in 2009

Perpetual Motion Robotics

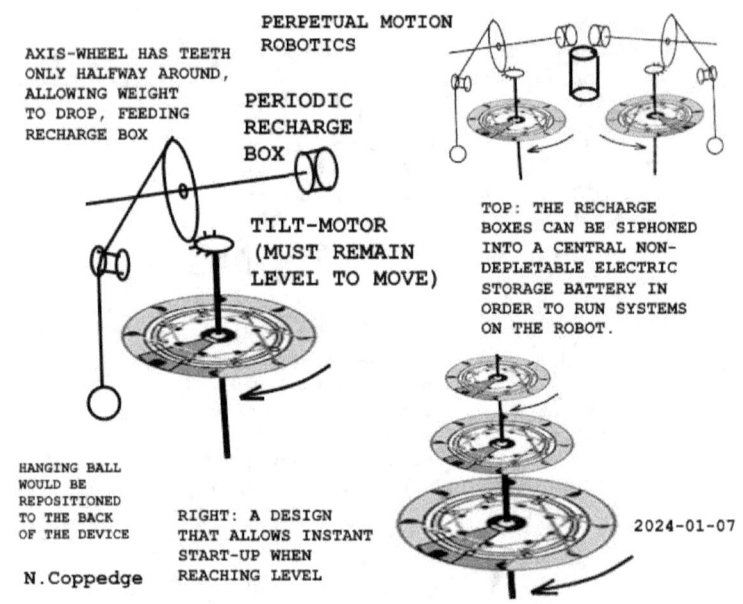

PERPETUAL MOTION ROBOTICS

AXIS-WHEEL HAS TEETH ONLY HALFWAY AROUND, ALLOWING WEIGHT TO DROP, FEEDING RECHARGE BOX

PERIODIC RECHARGE BOX

TILT-MOTOR (MUST REMAIN LEVEL TO MOVE)

TOP: THE RECHARGE BOXES CAN BE SIPHONED INTO A CENTRAL NON-DEPLETABLE ELECTRIC STORAGE BATTERY IN ORDER TO RUN SYSTEMS ON THE ROBOT.

HANGING BALL WOULD BE REPOSITIONED TO THE BACK OF THE DEVICE

N. Coppedge

RIGHT: A DESIGN THAT ALLOWS INSTANT START-UP WHEN REACHING LEVEL

2024-01-07

Pulley-Triangle

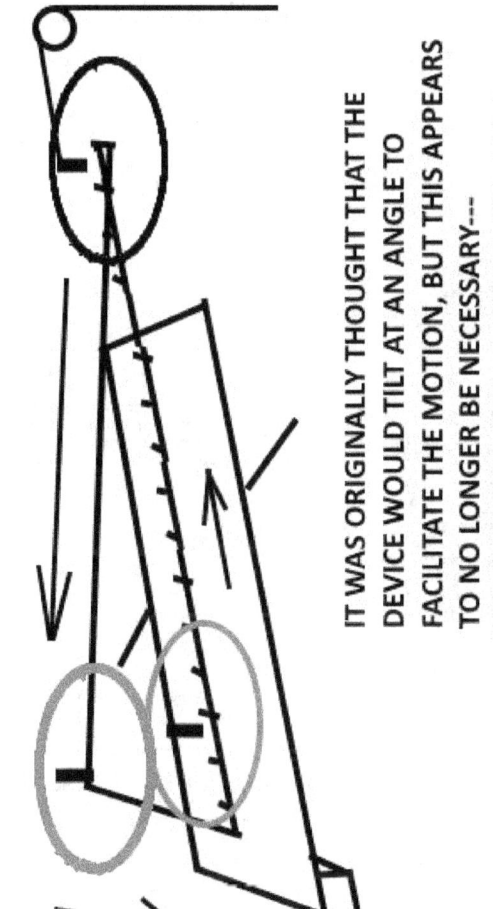

FIND A WAY TO ALLOW THE PULLEY WEIGHT TOO DROP AT AN ANGLE UNSUPPORTED, AND THIS DEVICE BECOMES AN INGENIOUS PERPETUAL MOTION MACHINE

IT IS A PROBLEM FIRST ENCOUNTERED WITH THE NIBW3

IT WAS ORIGINALLY THOUGHT THAT THE DEVICE WOULD TILT AT AN ANGLE TO FACILITATE THE MOTION, BUT THIS APPEARS TO NO LONGER BE NECESSARY----

THE KEY ELEMENT IS MAKING SURE THE SLIDING WEIGHT HAS LESS DOWNWARD RESISTANCE ON THE FURTHER SIDE.

Reactive Mechanisms

BASIC MECHANICAL REACTIONS

'MATH REACTION'

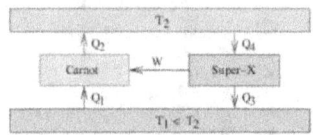

In the above diagram we can see the hypothetical 'super-x machine' as being equivalent to some special electronic process perhaps involving some type of overclocking or mathematical process using exponential efficiency.

'CAUSAL REACTION'

A series of elements, ideally each of them self-resetting, is set up so as to cause particular effects in each unit through an initial impulse. Mathematics or splash properties could be used in each unit so as to enhance the effect of the reaction.

'RISE AND SPLASH REACTION'

Using mechanical properties, energy might increase periodically, leading to a key major reaction using additional enhancement or chemistry. In this type, ideally the major reaction is charged periodically by the rising energy using mechanical properties or stored energy.

'SPLASH REACTION'

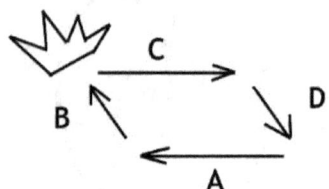

A basic chemical perpetual motion machine might not use a mathematical process, instead it might use something like flint-and-steel, with an object striking against a particular surface periodically. The type of reaction is not particularly important as long as it does not require much energy. What is important for the concept is that the cycle occurs periodically, involves some type of visible or audible reaction, and is repeatable. Probably the use of mechanical properties to create the reaction is ideal, in which case it is not inherently chemical at all. (For example, 'splash').

'ACCUMULATED REACTION'

Energy could be siphoned out of the reaction into an underlying field, which stores the energy, and feeds some quantity back into the system.

'MULTIPLIED REACTION'

Pictured at right, if a device is permitted to run in a channel, with other machines running parallel with the same direction of motion, this may be used to enhanced some of the other effects, and to create an enhanced reaction in the center of the path of motion.

Nathan Coppedge
2023-06-20
Publicly released.

MATH REACTION DIAGRAM BASED ON:
https://phys.libretexts.org/Bookshelves/University_Physics/Radically_Modern_Introductory_Physics_Text_II_(Raymond)/24%3A_The_Ideal_Gas_and_Heat_Engines/24.03%3A_Perpetual_Motion_Machines

Repeating Leverage 4.2 MODIFIED

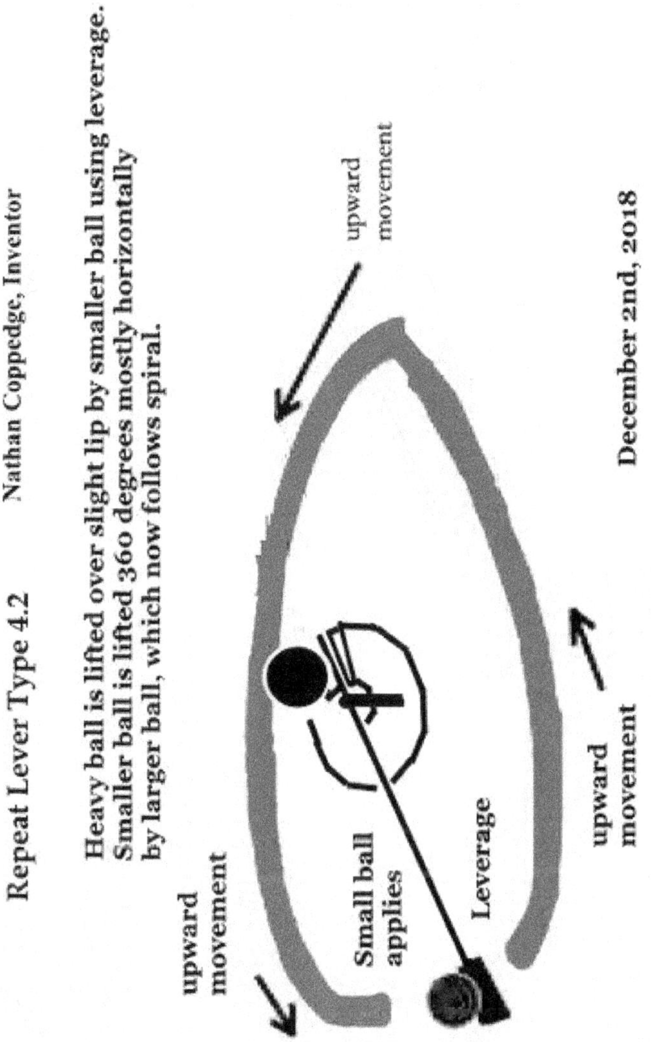

Repeat Lever Type 4.2   Nathan Coppedge, Inventor

Heavy ball is lifted over slight lip by smaller ball using leverage.
Smaller ball is lifted 360 degrees mostly horizontally
by larger ball, which now follows spiral.

upward
movement

upward
movement

upward
movement

Small ball
applies

Leverage

December 2nd, 2018

- Date of Invention: Dec 2, 2018.
- State of affairs: No known builds attempted yet.
- Rating: <150% conventional Over-Unity
- Leverage = 5:1
- Counterweight Mass = >3.5X to <6X
- Maximum Gradient: < 22.5 degrees.

Reverse RL 2, Improved

IMPROVED REVERSE RL2

Nathan Larkin Coppedge

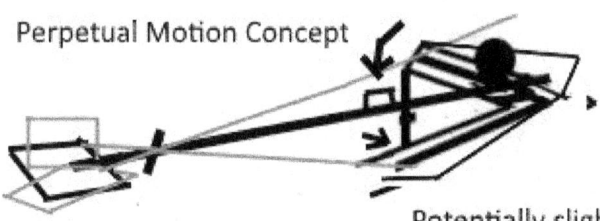

Perpetual Motion Concept

RATING: < 120%
LEVERAGE: 2 - 3: 1
COUNTERWEIGHT
MASS: >2.5X to <3X
EQUATION: STANDARD

Potentially slightly
tilted towards back,
including lever angle.

May 4, 2020

- Date of Invention: Potentially working design May 4, 2020.
- State of affairs: Still some doubts.
- Rating: < 120% conventional Over-Unity.
- Leverage: 2 - 3 : 1
- Counterweight Mass: 2.5X to 3X
- Maximum Gradient: < 9 degrees.

Reverse Swivel Device

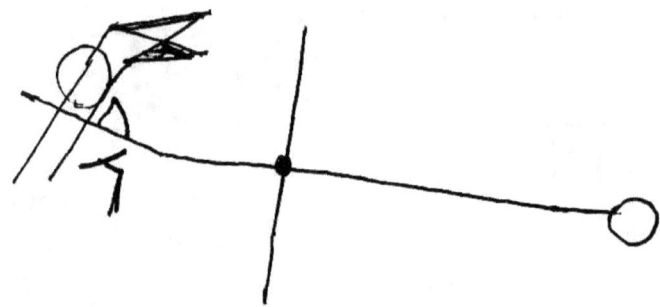

- Date of Invention: by August 8, 2019
- State of affairs: Save for later.
- Rating: Adjusted for larger weight < 150% formerly < 225% (Earlier less optimized: Adjusted < 123.5% formerly < 150% conventional Over-Unity)
- Leverage = 1:2 (1:1.5 is less optimized)
- Ball Mass = >2X to <3X (>1.75X to <2.5X is less optimized)

Scarpa's Pendulum

SCARPA'S PENDULUM

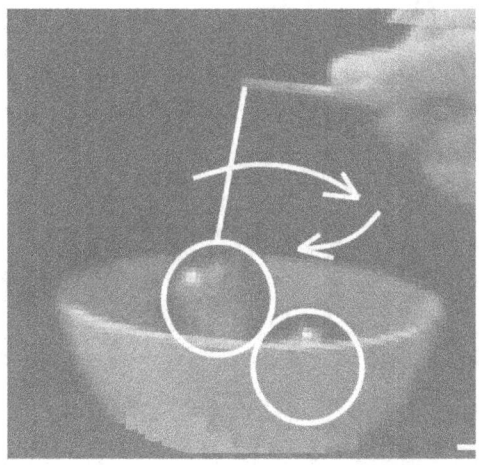

NATHAN LARKIN COPPEDGE

INSTRUCTIONS:
1. PLACE ONE BALL IN BAG.
2. TIE STRING TO STICK.
3. PLACE 2ND BALL IN BOWL
4. PLACE BAG IN BOWL AND ADJUST STICK UNTIL IT ROTATES.

LOOKS JUST LIKE PERPETUAL MOTION
!!

Self-Recharging Batteries

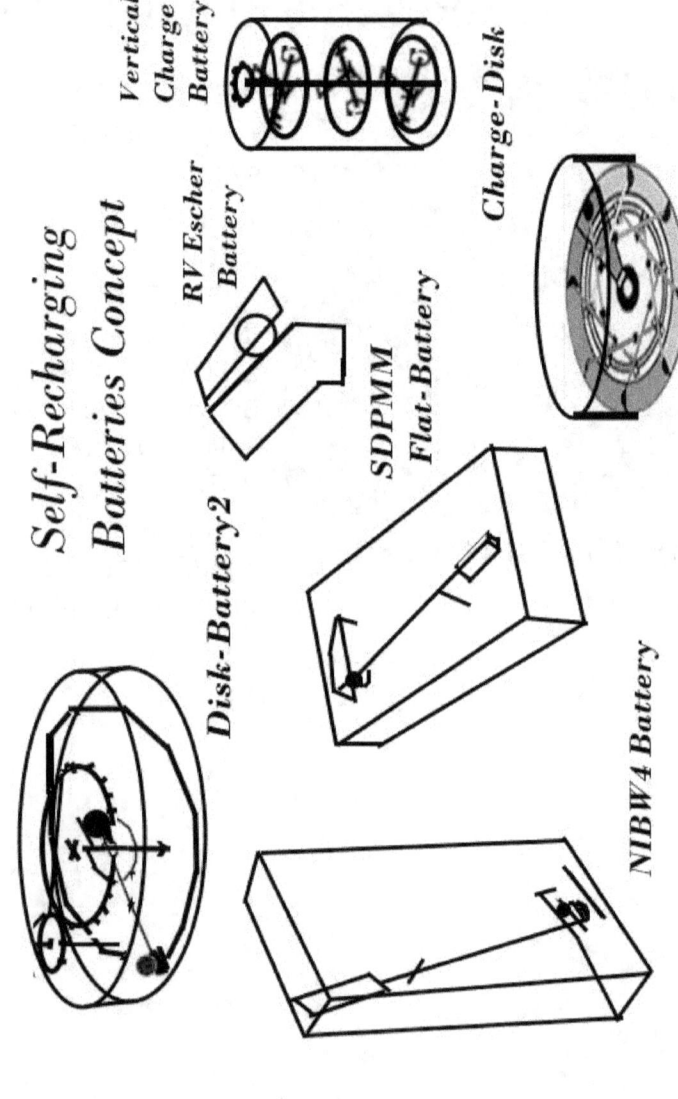

Self-Recharging Batteries Concept

Vertical Charge Battery

RV Escher Battery

Charge-Disk

SDPMM Flat-Battery

Disk-Battery2

NIBW4 Battery

Nathan Coppedge 2019 / 06 / 25

Self-Toggling Mechanism 1

Self-Toggling Mechanism 2

Shoes, Cat Spur (Safest Version 2023)

IMPROVED CATSPUR SHOES ITERATION 3

Nathan Coppedge
October 11, 2023

UPSIDE DOWN

THE RANGE IS PERMITTED TO BE SMALLER WHILE OFFERING LESS DANGER TO CHILDREN AND PETS BECAUSE OF THE INWARD MOTION.

RUBBER PLATE

HINGE

IN THIS CASE, WEDGE FORCE IS USED TO ENHANCE EFFECT OF LEVER ON PLATE WHILE CONVERSELY ALSO PERMITTING SHORTER MOTIONS OF LEVER.

NEW FEATURE DOESN'T GET STUCK ON MOST STAIRS

HINGE

REPULSION FORCE (KINETIC)

RUBBER PLATE UNDERNEATH SHOE

Shoes, Over-Unity

AUTHENTIC OU-SHOE

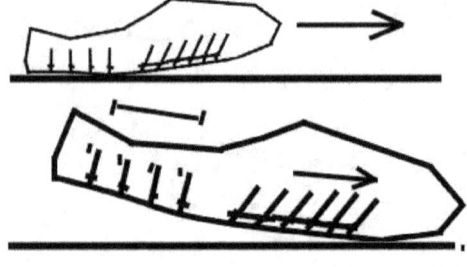

Nathan Coppedge
November 3, 2023

This uses small lightweight levers which are propelled forwards when the foot touches the ground using sliding material in base of shoe.

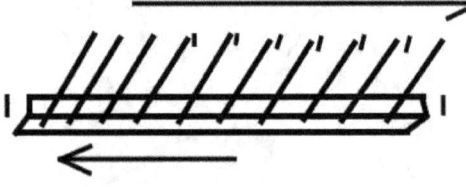

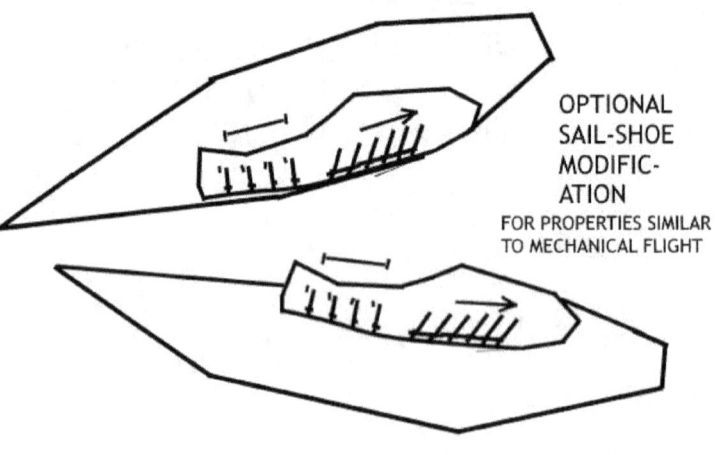

OPTIONAL
SAIL-SHOE
MODIFIC-
ATION

FOR PROPERTIES SIMILAR
TO MECHANICAL FLIGHT

Shoes, Wedge-Powered

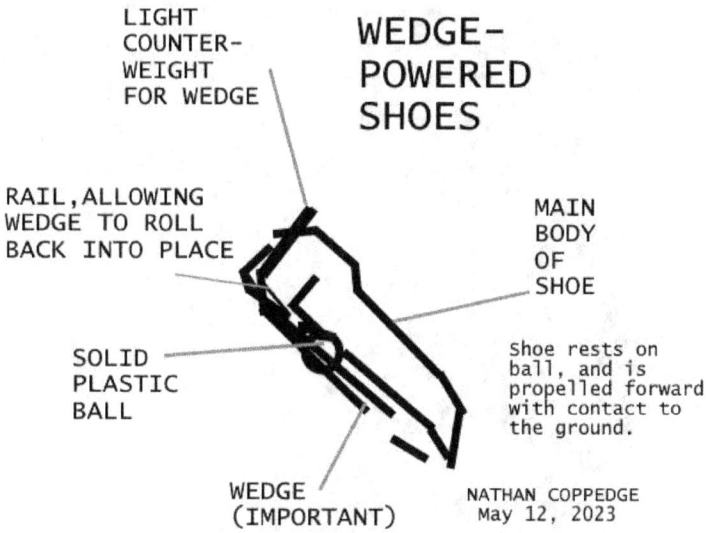

LIGHT
COUNTER-
WEIGHT
FOR WEDGE

WEDGE-
POWERED
SHOES

RAIL, ALLOWING
WEDGE TO ROLL
BACK INTO PLACE

MAIN
BODY
OF
SHOE

SOLID
PLASTIC
BALL

Shoe rests on
ball, and is
propelled forward
with contact to
the ground.

WEDGE
(IMPORTANT)

NATHAN COPPEDGE
May 12, 2023

Sideways-Swinging Vertical Lever Wheel

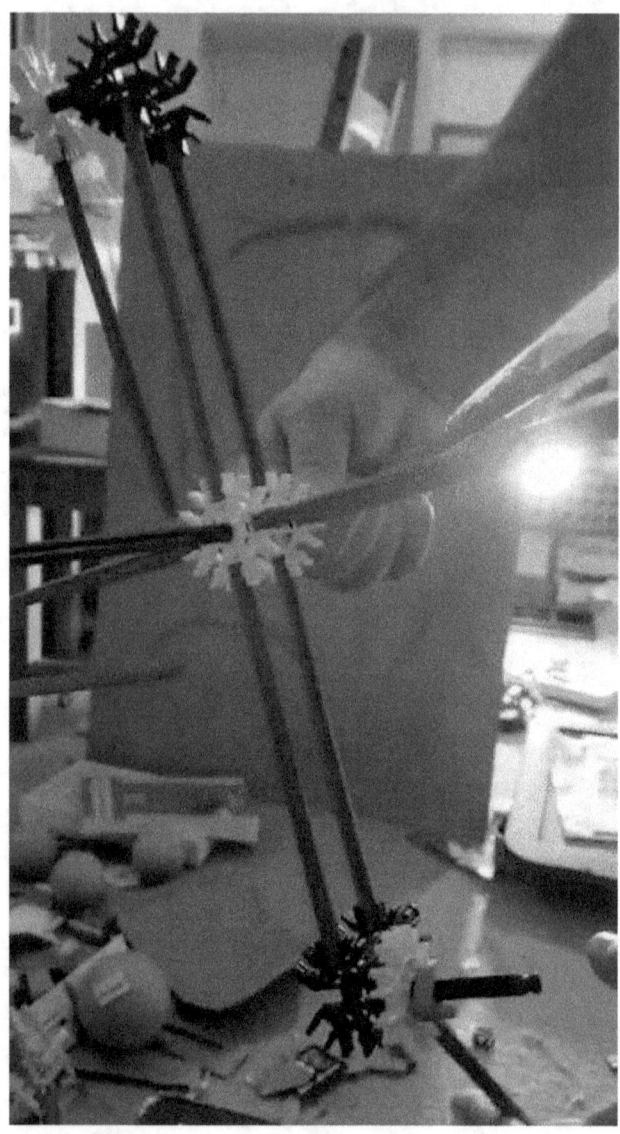

- Leverage: 2 : 1
- Counterweight: Roughly balanced mass.
- Rating: 125% (If there are two levers dropping at 2X torque with 0.5 mass on each of eight levers, then this increases the potential to: 2X Leverage X 2 X 0.5 + 1X leverage X 2 X 0.5 Vs 1.5 X 0.5 X 2 + 1 X 0.5 X 2 = (2X + 1X) : (1.5X + 1) = 3:2.5 = 125%)

Sideways Coquette

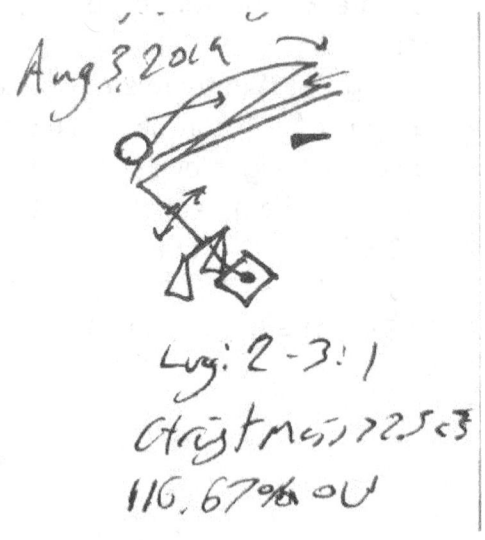

- Date of Invention: August 3, 2019
- State of affairs: May have a leverage problem, but a bit elegant, thought to work if re-designed.
- Rating: < 120% conventional Over-Unity.
- Leverage: 2 to 3 : 1
- Counterweight Mass = >2.5 <3
- Maximum Gradient: < 9 degrees.

Slanted Pulley Wheel

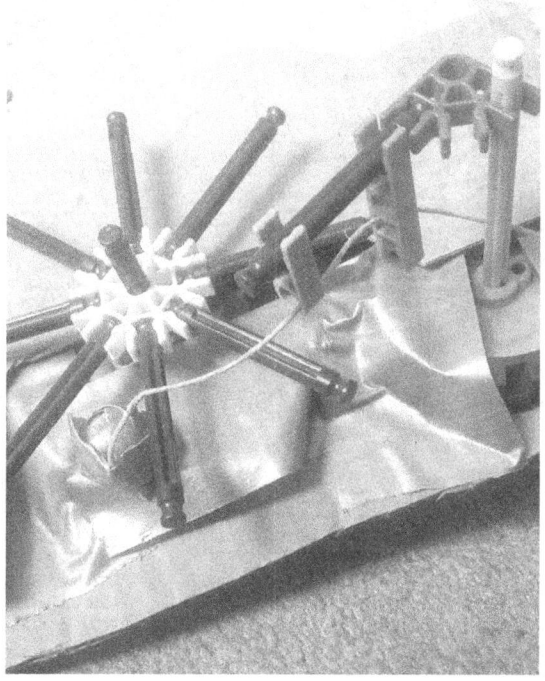

- Date of Invention: September 1, 2018 (?)
- State of affairs: May require additional cleverness.
- Rating: < 150% conventional Over-Unity.
- Leverage: 1 : 0.5
- Counterweight Mass: (1:1)
- Maximum Gradient: < 22.5 degrees.

Spindellabra Wheel

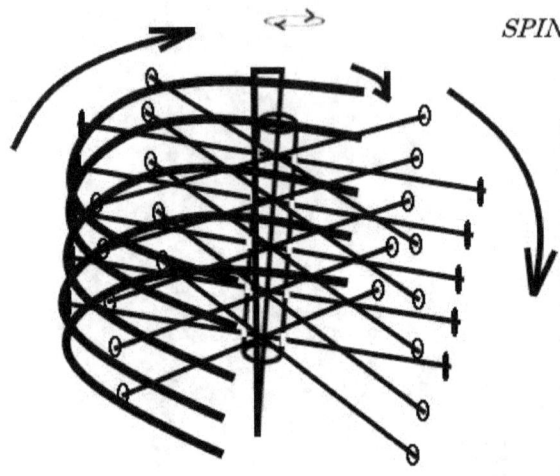

SPINDELLABRA WHEEL

WHEELS ARE BALANCED WITH SPOKES VERTICALLY-HINGED INDEPENDENT OF CENTRAL SPOKE WHICH ROTATES HORIZONTALLY, YET AT A VERY SLIGHT ANGLE AS SHOWN.

AT LEAST 2X WHEELS, NORMALLY BALANCED WITH THE OPPOSITES, FALL. HOWEVER, RISING WHEELS, NORMALLY AT LEAST 3X PER LEVEL, ARE SUPPORTED, REDUCING RESISTANCE ON RISING WHEELS BY APPROXIMATELY 50%.

3 X RISING X 0.51 = 1.53 X 5 = 7.65 EFFECTIVE LEVERAGE
2 X FALLING X 1 = 2 X 5 = 10

10 / 7.65 = APPROX 130.7% OVER-UNITY

TEST FOR DIFFERENCES WITH MORE LEVELS
1.53 X 10 = 15.3 EFFECTIVE LEVERAGE RISING
2 X 10 = 20 EFFECTIVE LEVERAGE FALLING
20 / 15.3 = SAME APPROX 130.7% OVER-UNITY
WITH MORE SPOKES.

EFFECT IS MAY BE EXAGGERATED BY THE EXISTENCE OF MULTIPLE LEVELS OF SPOKES, WHICH COMBINE WITH THE VERTICAL CENTRAL AXIS, WHICH AGAIN IS TILTED.

NATHAN LARKIN COPPEDGE 2023-11-21 NOTE THIS DEVICE BEARS RESEMBLANCE TO THE EARLIER DIABALANCE DEVICE ALSO KNOWN AS SWIVELING BALANCE, DATED TO JULY 19, 2021

Spiral Cone Device

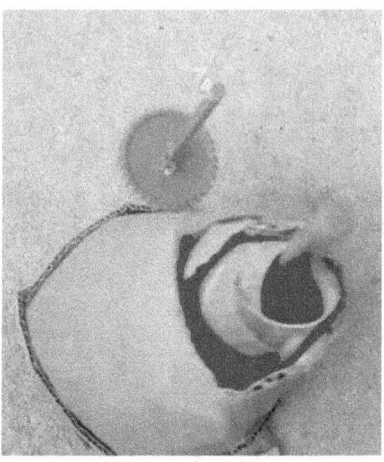

- Date of First Experiments: Sept 24, 2016.
- State of affairs: May have been replaced by the Differential Distance Pendulum. Requires more insight.
- Rating: <150% conventional Over-Unity (un-adjusted gives 150%, however all the stats match devices which have a rating of 150%).
- Leverage = 2 : 1 (However, a very short distance may be used. A rating of 1: 0.5 was avoided as I have found that rating makes calculation non-intuitive).

Spiral Pulley Lever Device

SPIRAL PULLEY LEVERAGE DEVICE

Assume 1 X counterweight, 1 X primary ball. 1 vs. 0.65 rising due to shallow slope of 0.65 X 1. Falling we can assume 1.5 X leverage with full application, 1.5 X 1 resistance from the ball

E. Advantage 1/0.65 one way, 1.5/1 the other.

Ball (A) begins at bottom Left of Spiral (B), progresses in a relatively narrow spiral upwards through action occurring through sideways pulley (C) attached by pin (D) and operated by counterweight (E). With ratio of approx 65% due to shallow upwards angle, motion continues. When Ball (A) reaches top of spiral, it is deflected sideways and outwards onto advantageous lever platform, returning backwards and deflecting inwards.

Nathan Larkin Coppedge   July 18, 2021

124

Date of First Invention: By August 1, 2018

Date of First Successful Diagram: July 18, 2021

Rating: < 120 % AVG

Leverage: 1 to 1.5 : 1

Counterweight Mass: 1X

Ball Range: > 1.75 < 2X

Maximum gradient: < 9 degrees.

Swapping Pulley Device

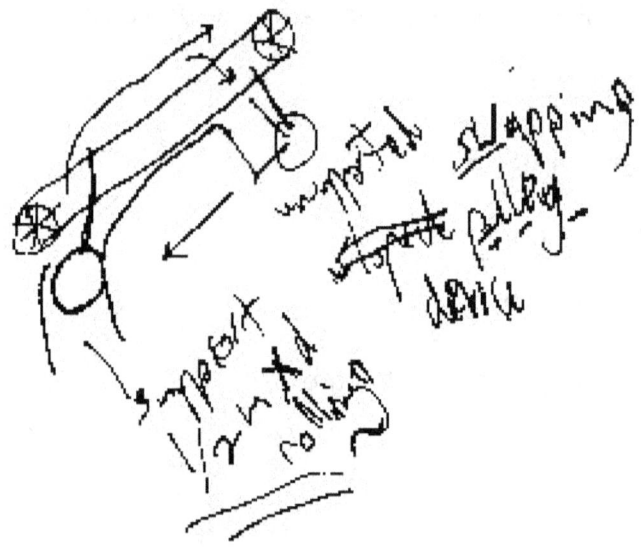

- NOTE: It was suspected that this design did not work, but it still seems that some effect could be created due to support versus non-support or even a wedge effect, although not quite as high as 150%. However, this design may also encounter an effect similar to balloons due to the vertical application of the weights and lack of leverage.
- Date of Invention: By August 14, 2019.
- Secretive clues: By December 29, 2016.
- State of affairs: Questions about falling weights being supported almost

horizontally. May require additional consideration even if workable.

- Rating: < 150% conventional Over-Unity.
- Simple Mass: 1:0.5
- Maximum Gradient: < 22.5 degrees.

Swiveling Balance Device

Two equal weights balance, the difference is that the low resistance from one weight due to a supporting track causes the device to swivel, allowing the other weight to return to the beginning. The device is designed so that the supporting track is slightly shorter than half of the overall rotation, thus the momentum of the upper weight can be used to swing the earlier weight into a position of support.

SWIVELING BALANCE DEVICE

Nathan Larkin Coppedge
July 19, 2021

First experiments: Similar devices were seen to climb steep surfaces in 2019.

Rating: < 125 % Conventional OU (Requires resolution of the transition problem, which is a hurdle that is generally cleared almost immediately before a device functions)

Leverage: 1 : 1 or equivalent

Ball Range: >2.5 < 3 (Note Difference of 2. The ball range is compared to the mass of one arm of the balance. Also requires flat and tilted trajectory for swivel)

Difference: 2 (the property of balance means in this case both sides have a weight advantage as one side receives support while the other side is dropping)

Maximum gradient: Unknown. Assumed steep.

Swivel Leverage Device

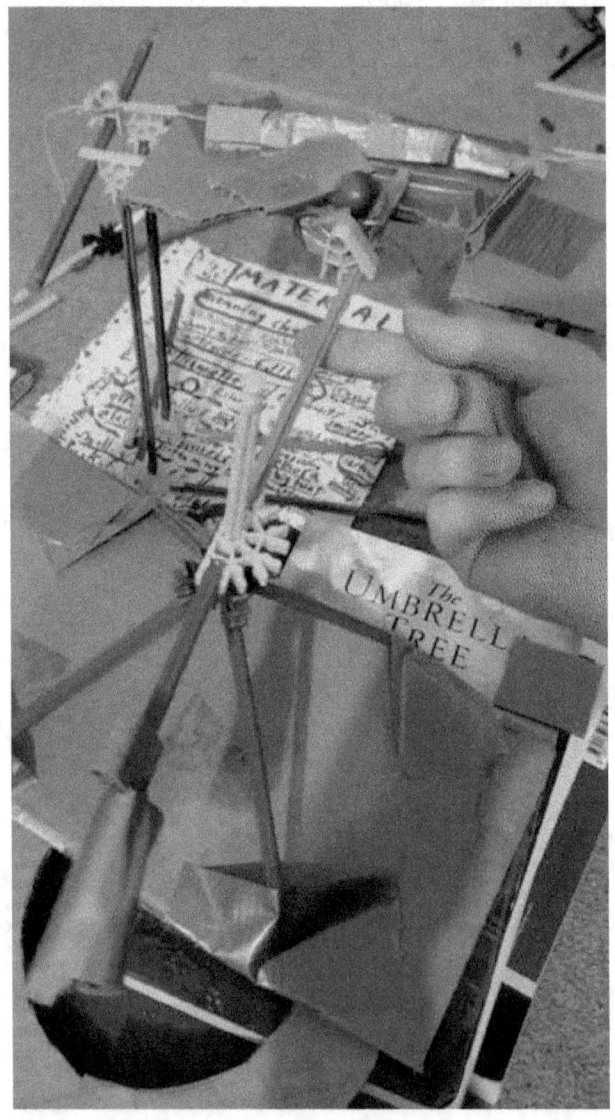

- Date of Invention: May 4, 2018 12 am.
- Date of First Partial Evidence: Oct 11, 2018.
- State of affairs: Some strong partial evidence.
- Rating: < 150% conventional Over-Unity
- Leverage = 2:1
- Counterweight Mass = >2X to <3X
- Maximum Gradient: < 22.5 degrees.

Tilt Motor

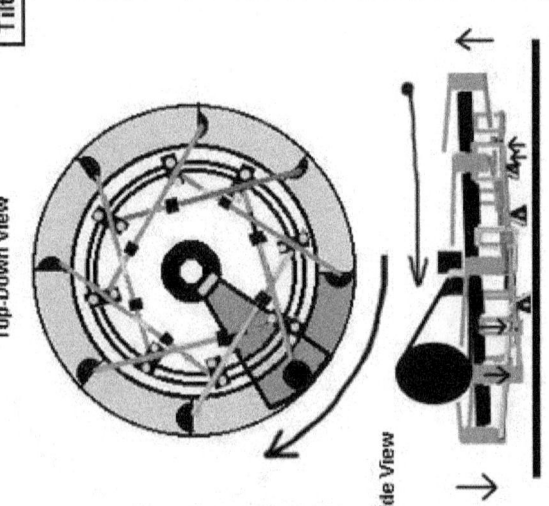

Tilt-Motor Perpetual Motion Concept

Original concept for a rotary device in which a weighted cone rolls around a swivel, activating successive pressure plates or "keys" operating levers. The levers in turn apply upwards pressure at a 90 degree angle on the track behind the cone. Since the track swivels downwards towards the portion weighted with the rolling cone, the upwards pressure is designed to create a continuous slope which follows the cone as it activates successive pressure plates.

Because the pressure plates are located outside the perimeter of the track, the cone's weight on the "wickets" on the active end of the levers only causes the pressure plate keys to be raised, rather than inhibiting movement by causing conflictive movement of the track. Metal "steps" are attached to the pressure plate keys in order to assure that the cone is activating one pressure plate at a given moment, which is meant to be sufficient to allow continuous motion.

Nathan L. Coppedge

Top-Down View

Side View

- Date of Invention: Oct 29, 2006.
- Date of First Evidence: April 2, 2007.
- Accurate Equations: After Aug 22, 2019.
- State of affairs: Not known if it has potential.
- Rating: < 150% conventional Over-Unity (4 mass range /(2 lvg * equivalent of 4 lever masses), calculate, + 1 * 100 = 150% OU).
- Leverage = 8/2X 2:1 lvg (Unconventional Extended Slope) alternately 1 : 0.5
- Rolling Cone Mass = >4X <8X of each counterweight mass (Fully updated. Formerly listed as: >1X and < 2X, and >2X Mass Resistance, this assumed totals of counterweight, which did not allow for correct leverage).
- Maximum Gradient: < 22.5 degrees.

Tilt Motor Car

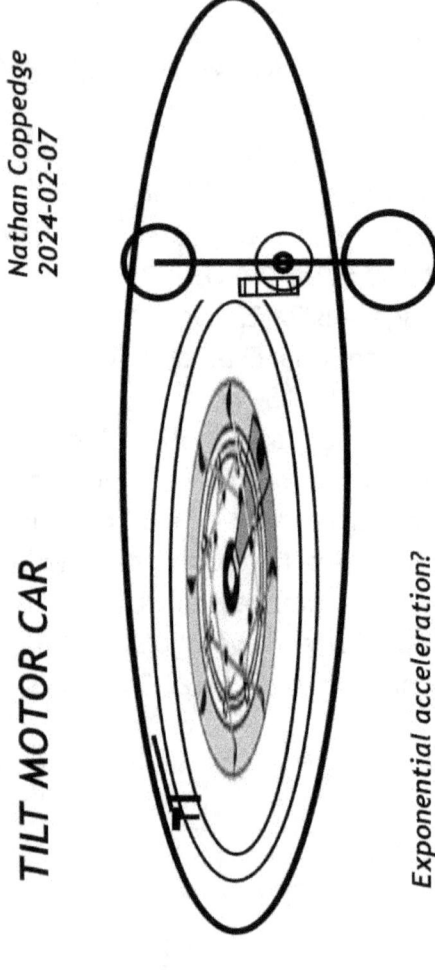

Nathan Coppedge
2024-02-07

TILT MOTOR CAR

Exponential acceleration?

"They move really fast. They look really big in the front. I don't even know how they move.... I don't know if we'll ever be enabled to build one of those things. I mean, it's so far in the future..."

Trapdoor Lever Device

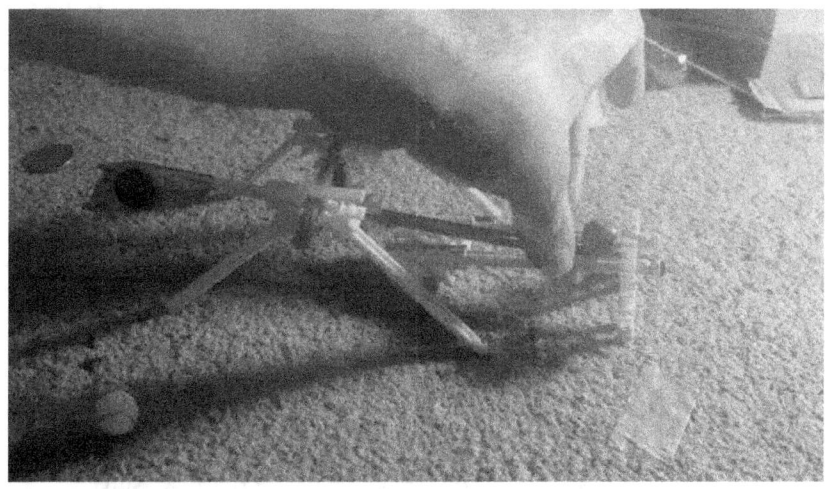

(Modular)

Leverage: 1.5 to 2 : 1

Counterweight mass: 2X - 2.5X smaller mass

Max OU Rating: < 125%

Smaller mass: 1X

Trough Lever

ARCHEMECHANICS: Trough Lever Device

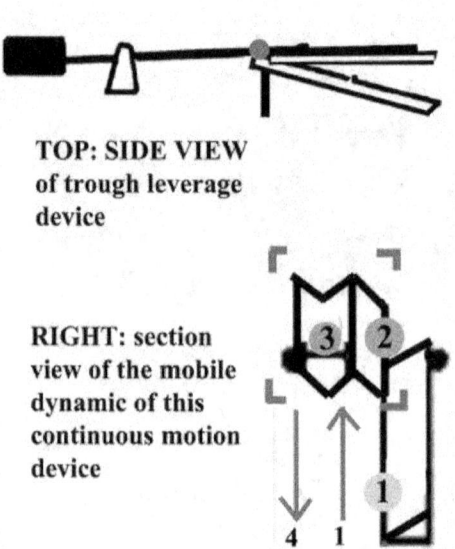

Double-trough device in which a partial trough assists upward leverage via a fixed half-track:

TOP: SIDE VIEW of trough leverage device

RIGHT: section view of the mobile dynamic of this continuous motion device

The dynamic is meant to promote energy generation by movement of a fulcrum, since greater weight bears on the mobile track when the ball weight is not supported by the fixed track segment

(c) 2011 Nathan Coppedge

- Date of Experiment: Dec 21, 2008.
- State of affairs: Similar to some possibly better designs, but may have design problems.
- <136% conventional Over-Unity.
- Leverage: 2.5–3:1

- Counterweight Mass = >2.5X to <3.5X
- Maximum Gradient: approx < 18 degrees (not calculated).

Unnatural Torque Device

Vertical Lever

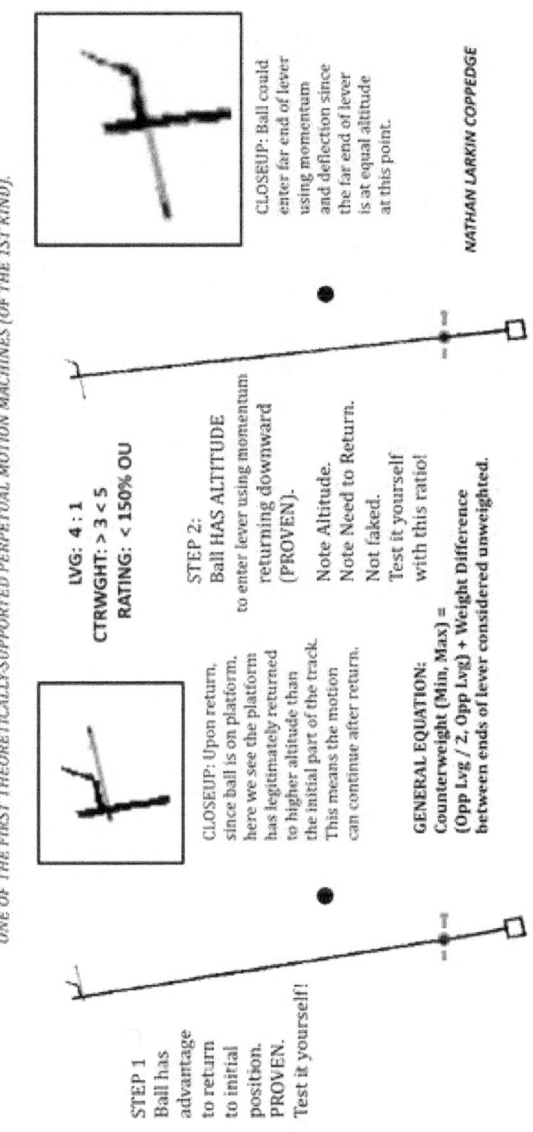

PROOF OF THE IMPROVED VERTICAL LEVER (IVL) PERPETUAL MOTION CONCEPT,
ONE OF THE FIRST THEORETICALLY-SUPPORTED PERPETUAL MOTION MACHINES (OF THE 1ST KIND).

CLOSEUP: Ball could enter far end of lever using momentum and deflection since the far end of lever is at equal altitude at this point.

NATHAN LARKIN COPPEDGE

LVG: 4 : 1
CTRWGHT: > 3 < 5
RATING: < 150% OU

STEP 2:
Ball HAS ALTITUDE to enter lever using momentum returning downward (PROVEN).

Note Altitude.
Note Need to Return.
Not faked.
Test it yourself with this ratio!

CLOSEUP: Upon return, since ball is on platform, here we see the platform has legitimately returned to higher altitude than the initial part of the track. This means the motion can continue after return.

GENERAL EQUATION:
Counterweight (Min, Max) = (Opp Lvg / 2, Opp Lvg) + Weight Difference between ends of lever considered unweighted.

STEP 1
Ball has advantage to return to initial position.
PROVEN.
Test it yourself!

- Date of Invention: September 27, 2019.
- Note: Some experimentation may be necessary in the angle of counterweight and distance of counterweight from the fulcrum ÷/- 1/16 of the lever's length, otherwise a flawless design
- State of affairs: No known builds. Thought to be one of the best and easiest designs.
- Rating: < 150% conventional Over-Unity.
- Leverage: 4:1 (looks like 8:1)
- Counterweight Mass: >3X to <5X
- Maximum Gradient: < 22.5 degrees.

Vertical NIBW5

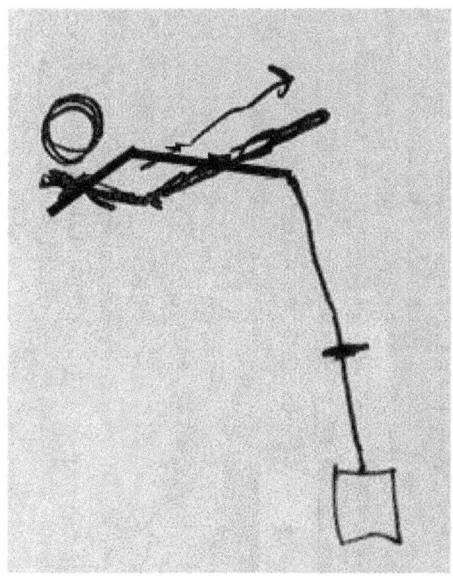

- Date of Invention: August 2, 2019.
- State of affairs: May be an improvement on the module for the typical NIBW 1, 5, however may be best if counterweight and lever is located horizontally from track to give ball more leverage.
- < 120% conventional Over-Unity.
- Leverage: 1 to 1.5:1
- Counterweight Mass: >1.75X <2X
- Maximum Gradient: < 9 degrees.

Vertical Wheel, Double Difference

VERTICAL WHEEL USING SPIRALS AND DOUBLE-DIFFERENCE

A vertical wheel made of spirals and joined from the exterior (not pictured), using two fixed horizontally rotating pendulums; The first [A] is heavier, acting on the second [B], which has enough weight to resist the wheel, providing upwards force against the first pendulum; The rotary motion of the first pendulum is meant to counteract the second, creating a circular motion of the wheel

[SIDE VIEW]

[FRONT VIEW]

Nathan Coppedge

SUGGESTED READING

Top 100 Ideas on How to Get Rich

50 Great Flying and Underwater Perpetual Motion Machines

The Black Swan Market

The Dimensional Philosopher's Toolkit

The Book of the Four

The Phenomenal Education

The Linguistics

Necessary Perfections

Bio

Nathan Coppedge or Nathan Larkin Coppedge (b.1982) is a philosopher, artist, inventor, poet, and member of the international honor society for philosophers. A prolific author with over 186 books published on Amazon, he is a perpetual motioneer, famous quotable, and internationally-selling Hyper-Cubist. A one-time member of Tesla Society UK online and PESWiki, and founder of many Facebook groups, he lives near Yale University.